T0074726

Phytochemistry of *Tinospora cordifolia*

Phytochemical Investigations of Medicinal Plants

Series Editor:
Brijesh Kumar

Phytochemistry of Plants of Genus *Phyllanthus*
Brijesh Kumar, Sunil Kumar and K. P. Madhusudanan

Phytochemistry of Plants of Genus *Ocimum*
Brijesh Kumar, Vikas Bajpai, Surabhi Tiwari and Renu Pandey

Phytochemistry of Plants of Genus *Piper*
Brijesh Kumar, Surabhi Tiwari, Vikas Bajpai and Bikarma Singh

Phytochemistry of *Tinospora cordifolia*
Brijesh Kumar, Vikas Bajpai and Nikhil Kumar

Phytochemistry of Plants of Genus *Rauvolfia*
Brijesh Kumar, Sunil Kumar, Vikas Bajpai and K. P. Madhusudanan

Phytochemistry of *Piper betle* Landraces
Vikas Bajpai, Nikhil Kumar and Brijesh Kumar

For more information about this series, please visit: https://www.crcpress.com/Phytochemical-Investigations-of-Medicinal-Plants/book-series/PHYTO

Phytochemistry of *Tinospora cordifolia*

Brijesh Kumar, Vikas Bajpai and
Nikhil Kumar

CRC Press is an imprint of the
Taylor & Francis Group, an **informa** business

First edition published 2020
by CRC Press
6000 Broken Sound Parkway NW, Suite 300, Boca Raton, FL 33487-2742

and by CRC Press
2 Park Square, Milton Park, Abingdon, Oxon, OX14 4RN

ISBN: 978-0-367-85964-0 (hbk)
ISBN: 978-1-003-01603-8 (ebk)

Typeset in Times
by codeMantra

Contents

List of Figures

List of Tables

List of Schemes

Preface

The scope of traditional medicinal herbs and their preparations has widely increased worldwide in recent years for the treatment of a range of diseases/ disorders. Identification and determination of the active phytochemicals are the crucial steps in the development of modern evidence based herbal medicines. Standardization and quality control of herbal extracts is an essential part of investigations to establish their safety and efficacy. Quality control process in herbal medicines involves plant authentication (at genus, species, and subspecies level) and establishment of marker profile for phytochemical screening of plant extracts and identification of marker compounds, to distinguish adulterants if any. To achieve this, development of validated methods for comprehensive chemical characterization and quantitation of the principal constituents of the herbal medicines is of prime importance. The inherent diversity in the structure of phytochemicals in medicinal plants and their qualitative and quantitative phytochemical characterization is not an easy task and requires highly sensitive and reliable modern analytical techniques, methods, and expertise to do it.

One of the trends in modern analytical research is the development of new methodology, which must be simple and capable to screen reliably all the major phytoconstituents in the sample bypassing the process of sample preparation. Direct analysis in real time mass spectrometry (DART-MS) is one such technique that offers sample analysis without any extraction (sample preparation), and it can analyze phytochemicals in intact plant material. Another mass variant is mass spectrometry (MS) coupled with liquid chromatography (LC) that has emerged in recent years as one of the most favored techniques for the identification of medicinally active ingredients in complex mixtures of plant extracts. Mass spectrometers having highly sensitive mass analyzers such as time of flight (TOF) provide high resolution mass data and help in the identification of analytes differing in their exact masses. LC-MS offers high degree of selectivity and sensitivity with the separation power. It allows simultaneous separation, identification, and structural analysis of phytochemicals present at the sub-ppm level in complex matrices of plant extracts.

This book describes the plant *Tinospora cordifolia*; and deals with the development and validation of DART MS, high performance liquid chromatography–electrospray ionization–quadrupole time of flight tandem mass spectrometry (HPLC-ESI-QTOF-MS/MS), and ultra performance liquid chromatography–electrospray ionization–quadrupole linear ion trap tandem mass spectrometry (UPLC-ESI-QqQ_{LIT}-MS/MS) methods for qualitative and quantitative phytochemical analysis of *T. cordifolia* stem.

T. cordifolia plant, especially the stem, is an important ingredient of many herbal formulations, reported in Ayurveda and other indigenous systems in the east Asiatic region. Its stem is an official medicine as listed by the Ayurvedic Pharmacopoeia of India. Though this plant is dioecious, it was not recognized until recent years. Several metabolites including alkaloids, sesquiterpenes, and phytoecdysteroids were identified and characterized in the samples obtained from different geographical locations and growing seasons and also the male and female plants of *T. cordifolia*. The information presented in this book will provide an opportunity to the readers to revisit their observations about the herbal drugs. We wish that this book will stimulate further research in plant sciences, biological sciences, chemical sciences, and pharmacological and clinical studies, which will add value to medicine discovery and product development using plants. It is hoped that this book will attract students, researchers, faculties, and herbal drug manufacturers involved in drug discovery and development.

31/12/2019 **The Authors**

Acknowledgments

The completion of this book is an outcome of several in house and collaborative projects carried out in the Sophisticated Analytical Instrument Facility (SAIF) Division, CSIR-Central Drug Research Institute (CDRI), Lucknow, India. We are glad to express our gratitude to people who have been supportive to us at every step. We express our deep sense of gratitude to CSIR-CDRI, Lucknow, India, for support. We are also grateful to Dr. K. P. Madhusudanan who pioneered the mass spectrometry in CSIR-CDRI, whose encouragement, guidance, and support from the beginning to the final stage enabled us to develop understanding of the subject. His constructive criticism and warm encouragement made it possible for us to bring the work to its present shape. We are indebted to him for shaping our thoughts. We are also thankful to all the team members and the director of CSIR-CDRI for their support during this period.

Authors

Dr. Brijesh Kumar is a Professor (AcSIR) and Chief Scientist of Sophisticated Analytical Instrument Facility (SAIF) Division, CSIR-Central Drug Research Institute (CDRI), Lucknow, India. Currently, he is facility in charge at SAIF Division of CSIR-CDRI. He has completed his PhD from CSIR-CDRI, Lucknow (Dr. R. M. L Avadh University, Faizabad, UP, India). He has to his credit 7 book chapters, 1 book, and 145 papers in reputed international journals. His current area of research includes applications of mass spectrometry (DART MS/QTOF LC-MS/4000QTrap LC-MS/Orbitrap MS^n) for qualitative and quantitative analyses of molecules for quality control and authentication/standardization of Indian medicinal plants/parts and their herbal formulations. He is also involved in identification of marker compounds using statistical software to check adulteration/substitution.

Dr. Vikas Bajpai completed his PhD from the Academy of Scientific and Innovative Research (AcSIR), New Delhi, India, and carried out his research work under supervision of Dr. Brijesh Kumar at CSIR-CDRI, Lucknow, India. His research interest includes the development and validation of LC-MS/MS methods for qualitative and quantitative analysis of phytochemicals in Indian medicinal plants.

Dr. Nikhil Kumar is a plant physiologist by training and worked in different capacities for about 30 years and finally superannuated from CSIR-National Botanical Research Institute (NBRI), Lucknow, India. He has worked on one of the very important cultural plants of India including Southeast Asia. *Piper betle* effectively broadened his understanding as how plants contributed in the development of human skills. He has published more than fifty research papers in national and international journals. He brought to focus the aspect of dioecy in *Piper betle* and *Tinospora cordifolia* and their possible functional implications in adaptation and biological activities.

List of Abbreviations and Units

°C	degree Celsius
μg	microgram
μL	microliter
APCI	atmospheric pressure chemical ionization
API	atmospheric pressure ionization
BPC	base peak chromatogram
CAD	collision activated dissociation
CE	capillary electrophoresis
CE	collision energy
CID	collision induced dissociation
CXP	cell exit potential
Da	Dalton
DAD	diode array detection
DART	direct analysis in real time
DP	declustering potential
EP	entrance potential
ESI	electrospray ionization
FDA	Food and Drug Administration
FIA	flow injection analysis
g	gram
GC-MS	gas chromatography–mass spectrometry
GS1	nebulizer gas
GS2	heater gas
h	hour
HPLC	high performance liquid chromatography
ICH	International Conference on Harmonization
IS	internal standard
IT	ion trap
kPa	kilopascal
L	liter
LC	liquid chromatography

LOD	limit of detection
LOQ	limit of quantification
LTQ	linear trap quadrupole
m/z	mass to charge ratio
mg	milligram
min	minute
mL	milliliter
mM	millimolar
MRM	multiple reaction monitoring
MS	mass spectrometry
ms	milli second
MS/MS	tandem mass spectrometry
ng	nanogram
NMPB	National Medicinal Plants Board
NMR	nuclear magnetic resonance
PCA	principal component analysis
PDA	photo diode array
psi	pressure per square inch
QqQ_{LIT}	hybrid linear ion trap triple quadrupole
QTOF	quadrupole time of flight
r^2	correlation coefficient
RDA	retro-Diels Alder
RSD	relative standard deviation
S/N	signal-to-noise ratio
SD	standard deviation
TC	*Tinospora cordifolia*
TCS	*Tinospora cordifolia* stem
t_R	retention time
TSM	traditional system of medicine
UHPLC	ultrahigh performance liquid chromatography
UV	ultraviolet
WHO	World Health Organization
XIC/EIC	extracted ion chromatogram

Introduction *Tinospora cordifolia* (*Amrita*)—The Wonder Plant

1

Tinospora cordifolia Linn. (*Amrita*) is just 1 out of 3,69,000 known species of vascular plants on the earth. It is interesting that only 31,128 species out of this big pool are with known uses (fulfilling human, cattle, or environmental needs) and just 103 species meet more than 90% human food requirements (Bramwell, 2002). The largest subgroup of the useful plants numbering more than 17,000 species contribute to the human wellness and are collectively known as medicinal plants. *T. cordifolia* (TC) belongs to this exclusive group. Thus, in the ancient societies or even much before their establishment, humans had well developed skills of exploration, observation, and methods to pass on the information to later generations. This is a testimony to the human ability to find a needle in a haystack! The fact that a large number of the world's population still depends on the medicinal plants (WHO, 2002) through the traditional system of medicine (TSM), underscores their importance even in the present times when synthetic or biosimilar modern drugs are there. Currently, many plant species are under threat of extinction due to the overexploitation and pressure on their habitats, and the looming danger of climate change will also negatively impact TSM and also the population dependent on it (Chen et al., 2016).

1.1 *NATURAL HABITAT AND DISTRIBUTION*

The genus *Tinospora* belongs to the family Menispermaceae, which has 13 accepted species even though more than 65 species have been reported. Members of this family are widely used in TSM due to their medicinal properties and rank second only to the great neem *Azadirachta indica* known globally for its medicinal uses. In India, genus *Tinospora* is represented by four species: *Tinospora cordifolia* and *Tinospora sinensis* (syn. *Tinospora malabarica* Miers ex Hook. f.) are widely distributed, and *Tinospora crispa* and *Tinospora glabra* are reported only from Northeast India and the Andaman Islands (Pramanik et al., 1993). TC is a very widely used plant in TSM. The occurrence of the genus *Tinospora* has been reported from Africa, Bangladesh, Cambodia, China, India, Indonesia, Myanmar, Pakistan, Philippines, Thailand, and Vietnam (Figure 1.1) showing a wide range of distribution and TC has been also reported from most of these regions.

In India, the plant was first botanically identified in the year 1806 during botanical expeditions. It has been reported from semi arid, tropical and subtropical regions; its occurrence has been also reported from the western Himalayas up to 1,000 M above mean sea level (Rana et al., 2012; Singh et al., 2014; Singh and Thakur, 2014). In India cultural integration of TC is far and wide in the regions of its occurrence and use. There are more than a hundred names given to this plant showing its reach within the population. The Sanskrit

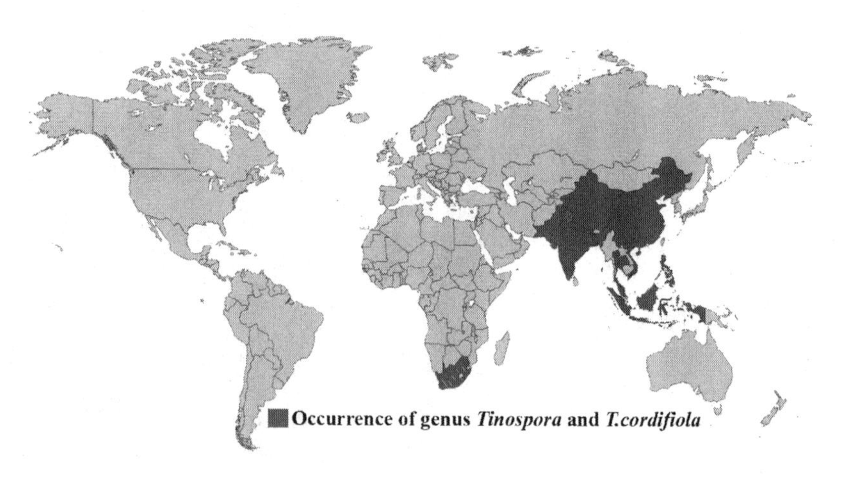

FIGURE 1.1 Worldwide distribution of genus *Tinospora* and *Tinospora cordifolia*.

names such as *Amrita* (nectar) and *Guduchi* (the one who protects) are more common and widely used.

Amrita/Guduchi is one of the highly revered herbs of Ayurvedic medicines. The plant has also different local names in different countries like in Indonesia, it is Brotowali, Andawali, and Putrawali; in Philipines, it is Makabuhay (Tagolog), Paliaban (Bisaya), Pañgiauban (Bisaya), and Taganagtagua; in Thailand, it is Boraphet; in China, it is K'uan Chu Hsing; and in Myanmar, it is SIndone-ma-new. The common English name is heartleaf moonseed based on the shape of the leaf, which eventually contributed to the nomenclature "*cordifolia*." *Amrita* vines along with bilva (*Aegle marmelos*), bhang (*Cannabis indica*), tulsi (*Ocimum sanctum*), and lotus flowers (*Nelumbo nucifera*) are offered to Lord Shiva. The following quote from Charak Samhita describes some of its properties:

अमृता सांग्रहिकवातहर दीपनीय श्लेस्माषोणित विबन्धप्रशमनानां (Charak samhita sootrsthaan 25).

Meaning the plant compacts (compress), suppresses vata, stimulates appetite, cures cough, purifies blood, and removes constipation.

रसायन नाम तद ज्ञेयं यद्‌ जरा व्याधि नाशनं ! यथा अमृता.!! (Sharangdhar Samhita 4/14 Murthi 1984).

Rasayanas are those plants (drugs) that restrict aging (general decline) and diseases; one of the examples is *Amrita*.

1.1.1 The Plant

TC is a perennial, vigorous, softwood climber with no special features for this habit except the winding flexibility of the stem to wind clockwise around the support, mostly trees, shrubs, or any other natural support or provided anthropogenically (Choudhary et al., 2014). Such types of climbers are also known as lianas, which are important part of the forest ecosystems. The plant is deciduous; in subtropical India, the leaf fall occurs with the onset of winter (mid-October to end of November), and new leaves appear in the spring (February). Flowering starts with the onset of leaf fall and continues until April to May. In some other locations (Bangalore) with a mild climate, flowering continues from August to January. TC is a dioecious plant having male (staminate) and female (pistillate) flowers on different plants. Pollination is entomophilous, i.e., by insects, and the plants bear small berry like fruits in bunches similar to grapes. Unripe fruits are green, and when ripe, the fruits get bright red pigmentation and appear quite attractive. Ripe fruits are bitter sweet in taste and eaten by birds, which also helps in propagation. Since the plant has tremendous regeneration capacity, vegetative propagation is the main route to spread. All parts of the plant are used as a drug, but the stem is preferred over the rest.

1.2 MICROSCOPY OF *TINOSPORA CORDIFOLIA* STEM

Central portion of stem is occupied by parenchymatous pith in young stem. Starch grains are abundant in all the parenchymatous cells. Mucilage canals are also found scattered in cortical, phloem, and pith regions (Choudhary et al., 2014).

1.3 AYURVEDIC PROPERTIES

According to Ayurveda, the properties ascribed to TC are rasa (taste), tikta (bitter), kashaya (astringent), guru (heavy), snigdha (unctuous), veerya (potency), ushna (warm), vipaka (post digestive effect), and madhura (sweet). The stem is primarily made up of thin walled cells, which store the matabolites used as drugs. The stem as natural sink has highest dry matter compared to leaf, flower, and fruits making it the most suitable source of drug.

In synthetic terms, the stem is bitter, astringent, sweet, thermogenic, anodyne, anthelmintic, alternate, antiperiodic, appetizer, antispasmodic, antiinflammatory, antipyretic, digestive, carminative, appetizer, stomachic, constipating, cardiotonic, depurative, hematinic, expectorant, aphrodisiac, rejuvenating, galactopurifier, and tonic. It is useful in vitiated conditions of *vata*, burning sensation, hyperdipsia, helminthiasis, dyspepsia, flatulence, stomachalgia, intermittent fevers, chronic fevers, inflammation, gout, vomiting, cardiac debility, skin diseases, leprosy, erysipelas, anemia, cough, general debility, jaundice, seminal weakness, uropathy, and splenopathy (Warrier et al., 1993).

All the important medicinal plants have associated folklores/tales so is the case with TC. The sacred origin of *Amrita/Guduchi* is described in the Indian epics *Ramayana* and *Durga Saptshati*. The *Ramayana* states that when Lord Rama made a special prayer to Lord Indra to resurrect all the monkeys and bears of his army killed in the war with the demon king, Ravana, Lord Indra yielded to his prayers' and said let that be and sprinkled nectar from the heavens to resurrect the animals. As the nectarous drops fell upon the dead, they suddenly came back to life. Some of the nectar drops that also fell on the earth formed the sacred *Amrita/Guduchi* plant. *Amrita* is one of the Rasayana plants of Ayurveda extensively used in a score of ailments. It is one of the Vedic plants that also figures in different texts of Ayurveda such as *Charak Samhita*, *Shusrut Samhita*, and *Vrahattrayi* (Upadhyay et al., 2010) and its use may be as old as 5000 BC., *Amrita* is also considered to be the soma of *Rigveda*. Since this plant belongs to the special category, Rasayana, several functions have been ascribed to TC, which are listed in Table 1.1.

Rasayana is one of the eight clinical specialties of classical Ayurveda. Rasayana replenishes/balances the vital body fluids and protects from diseases. Rasayanas are rejuvenators and nutritional supplements and possess strong antioxidant activity (Chulet and Pradhan, 2009). It also pacifies all *tridoshas* (vat, kaph, and pitta) and maintains their balance with each other, thus imparting it Rasayana character. The diseases have their origin in imbalances of the three basic rasas: vata, pitta, and kapha. It balances vata when taken with ghrita (clarified butter), pitta when taken with *guda* (jaggery, concentrated sugarcane juice), and kapha when taken with honey.

TC stem finds a special mention for its use in tribal or folk medicine in different parts of India. TC is reported as a tonic for treatment of allergies, diabetes, dysentery, jaundice, heart diseases, leprosy, metabolic disorders, rheumatoid arthritis, skin diseases, and urinary disorders in Ayurveda (Gupta et al., 2011). *Guduchi* satva (also known as *Giloi starch*), a kind of starch obtained from processed aqueous extract of dry stem powder, is widely used.

TABLE 1.1 Ayurvedic terms depicting *Tinospora cordifolia* functions

S. NO.	*AYURVEDIC TERMINOLOGY (SANSKRIT)*	*WORKING MEANING*
1.	Vaya sthapana	Prolongs longevity
2.	Daha prashaman	Antiinflammatory
3.	Trishna nigrahana	Reduces feeling of excessive thirst (throat drying)
4.	Stanya shodhana	Increases lactation and quality
5.	Triptighna	Increases appetite and reduces anorexia
6.	Rasayani	Rejuvenator
7.	Samgarhini	Promotes absorption (liquid from Malas)
8.	Balya	Increases strength
9.	Agani deepani	Ignites gastric fire
10.	Valee palita nashini	Delays appearance of aging signs
11.	Medhya	Enhances memory and grasping power

1.4 GROSS PHARMACOLOGICAL PROPERTIES

Investigation on this plant has indelible Indian signatures as the first publication appeared in *Indian Journal of Medical Research* by Pendse and Dutt in 1932 and also the second paper by Bhide et al. (1941); however, the details are not available. The first documented antiinflammatory effect was reported by Pendse et al. (1977). Based on Pubmed, the take off year of publication based on TC is around 1986, and papers started appearing with upward trend and the jump registered around 2010, which is still on. It is also noteworthy that majority of the findings have appeared from India. The advancements in instrumentations enabling analysis based on crude extracts or even whole tissue/organ have further given a new boost to (re)visit the plants as major resource for human wellness.

Pendse et al. (1977) showed that the water extract of the TC stem growing on *neem* (*Azadirachta indica*) showed antiinflammatory, analgesic, and antipyretic actions in albino rats and immunosuppressive effect in albino rabbits. Antiallergic, antiinflammatory, and antileprotic properties were shown in stem bark (Asthana et al., 2001). TC stem bark is reported to possess analgesic effect, and stem extract is found to be useful in treatment of skin diseases (Upadhyay et al., 2010). Antidiabetic, cytotoxic, hepatoprotective, and immunomodulatory activities were observed in stem (Bishayi et al., 2002; Jagetia and Rao, 2006a). TC leaves exhibited antidiabetic action in alloxan rats and showed insulin like action (Sharma et al., 2015). Effectiveness of TC aqueous extract in treatment of urinary diseases and rheumatism was reported by Ahmed et al. (2015). TC stem is used in nutraceuticals as tonic (Gupta et al., 2011) due to its antiinflammatory, antiperiodic, antipyretic, and antispasmodic effect. The TC fruit is also observed to be active in treatment of jaundice and rheumatism (Upadhyay et al., 2010). TC roots are potent emetic agent and are useful as antistress, antiulcer, hypoglycemic, and visceral obstruction conditions (Dwivedi and Enespa, 2016). The whole TC plant comprising leaf, stem, roots, flowers, and fruits (collectively known as *panchang* in Ayurveda) possess antioxidant, antiulcer, and hepatoprotective activity (Saha and Ghosh, 2012). Kidney, liver, and spleen are the main target organ sites where TC activity is high (Gupta et al., 2011). Treatment in mice and rats with TC extract showed protection against induced infections, and it has been also found to be nontoxic in humans (Kalikar et al., 2008; Dwivedi and Enespa, 2016). The chemical constituents mainly alkaloids, bitter compounds, and lipids contribute to the medicinal effects of the plant (Reddy and Reddy, 2015).

The TC stem is the official Ayurvedic medicine listed in the Ayurvedic Pharmacopoeia of India (The Ayurvedic Pharmacopoeia of India, 1999;

Choudhary et al., 2014). TC is one of the most versatile rejuvenating herbs promoting longevity and hence also called Vayastha or *Amrita* (Khan et al., 2016). The whole plant is also used in veterinary, folk, and Ayurvedic system of medicine along with other plants in commercial formulation (Sharma et al., 2010a). Traditionally, *Guduchi satva* obtained from TC stem growing on neem tree is considered to be more efficacious; however, it needs validation.

1.5 PHYTOCHEMISTRY

A wide range of diverse compounds of different classes have been isolated from TC such as norclerodane diterpenoids, essential oils, fatty acids (Qudrat-I-Khuda et al., 1964; Khuda et al., 1966), furano diterpenoids (Kohno et al., 2002), phenolic lignans, polysaccharides (Leyon and Kuttan, 2004), alkaloids (Sarma et al., 2009), sterols, and terpenoids (Singh et al., 2017). Many alkaloids have been reported from this plant, such as berberine, palmatine, jatrorrhizine, tetrahydropalmatine, magnoflorine, and tembetarine (Sarma et al., 1998, Singh et al., 2017). Some aporphine glycoside alkaloids like tinocorside A, tinocorside B (Maurya et al., 1996; Maurya et al., 1998), and quaternary saturated amine choline (Bajpai et al., 2016; Singh et al., 2017) along with phenolic amide *N*-trans-feruloyltyramine (Bajpai et al., 2016) were also reported from the stem. Several clerodane, norclerodane diterpenoids including their glycosides are the main chemical constituents reported in the stem. Presence of monoterpenoids, sesquiterpenoids, and triterpenoids (cycloeuphordenol) were also reported from the aerial parts (Maurya et al., 1998; Sharma et al., 2010a, 2010b; Singh et al., 2017). Other reported compounds included phenyl propanoids (syringin and cordioside D), lignans (pinoresinol-di-O-glucoside, secoisolariciresinol-9′-O-β-D-glucopyranoside, and 3-(α, 4-dihydroxy-3-methoxybenzyl)-4-(4-hydroxy-3-methoxybenzyl)tetrahydrofuran along with phenolic compounds such as 2-hydroxy-3-methoxy benzaldehyde and 4-hydroxy-3-methoxy benzoic acid (Maurya et al., 1998; Singh et al., 2017), (–) epicatechin (Pushp et al., 2013; Singh et al., 2017), and steroids such as δ-sitosterol, 2,3,14,20,22,25-hexahydroxyl-5-cholest-7-en-6-one from stem (Singh et al., 2003a). Aliphatic compounds such as methyl-9,12-octadecadienoate, methyl-9-octadecenoate, methyl hexadecanoate, and methyl octadecanoate (Sharma et al., 2010b; Singh et al., 2017), polysaccharides composed of glucose, arabinose, rhamnose, xylose, and mannose (Jahfar, 2003; Singh et al., 2017), and G1-4A, an arabinogalactan polysaccharide and α-D-glucan, were also identified and isolated from this plant. Major compounds reported from TC plant are shown in Table 1.2.

TABLE 1.2 Major phytochemicals isolated and identified from different plant parts of *Tinospora cordifolia*

S. NO.	*COMPOUNDS*	*REPORTED PART*	*REFERENCES*
1.	10-Hydroxycolumbin	Whole plant	Sivasubramanian et al., 2013
2.	18-Nonderodane glycoside	Whole plant	Maurya et al., 1998
3.	1-Hexanol	Leaf	Naik et al., 2014
4.	1-Penten-3-ol	Leaf	Naik et al., 2014
5.	2, 4-Pentadienal	Leaf	Naik et al., 2014
6.	2-Hexen-1-ol	Leaf	Naik et al., 2014
7.	2-Hexenal	Leaf	Naik et al., 2014
8.	2-Penten-1-ol	Leaf	Naik et al., 2014
9.	2-Phenoxy ethanol	Leaf	Naik et al., 2014
10.	3-Hexen-1-ol	Leaf	Naik et al., 2014
11.	4-Hexen-1-ol	Leaf	Naik et al., 2014
12.	8-Hydroxycolumbin	Whole plant	Sivasubramanian et al., 2013
13.	8-Hydroxytinosporide	Whole plant	Sivasubramanian et al., 2013
14.	Berberine	Whole plant	Panchabhai et al., 2008
15.	Biphenyl	Leaf	Naik et al., 2014
16.	Choline	Whole plant	Panchabhai et al., 2008; Bajpai et al., 2016
17.	Clerodane derivatives	Whole plant	Panchabhai et al., 2008
18.	Columbin	Whole plant	Sarma et al., 1998; Panchabhai et al., 2008; Sivasubramanian et al., 2013; Singh and Chaudhuri, 2015
19.	Cordifelone	Whole plant	Panchabhai et al., 2008
20.	Cordifol	Whole plant	Panchabhai et al., 2008
21.	Cordifolide A	Stem	Pan et al., 2012
22.	Cordifolide B	Stem	Pan et al., 2012
23.	Cordifolide C	Stem	Pan et al., 2012
24.	Cordifolioside A	Whole plant	Panchabhai et al., 2008
25.	Cordifolioside B	Whole plant	Panchabhai et al., 2008
26.	Cordifolioside C	Whole plant	Panchabhai et al., 2008

(*Continued*)

TABLE 1.2 (*Continued*) Major phytochemicals isolated and identified from different plant parts of *Tinospora cordifolia*

S. NO.	*COMPOUNDS*	*REPORTED PART*	*REFERENCES*
27.	Cordifolioside D	Whole plant	Panchabhai et al., 2008
28.	Cordifoliside A	Whole plant	Panchabhai et al., 2008
29.	Cordifoliside B	Whole plant	Panchabhai et al., 2008
30.	Cordifoliside C	Whole plant	Panchabhai et al., 2008
31.	Cordifoliside D	Whole plant	Panchabhai et al., 2008
32.	Cordifoliside E	Whole plant	Panchabhai et al., 2008
33.	Cordioside	Whole plant	Panchabhai et al., 2008
34.	Corydine	Stem	Singh and Chaudhuri, 2015
35.	Di-*N*-octyl phthalate	Leaf	Naik et al., 2014
36.	Diterpenoid	Whole plant	Panchabhai et al., 2008
37.	Ecdysterone	Whole plant	Panchabhai et al., 2008
38.	Furanoid diterpene glycoside	Whole plant	Panchabhai et al., 2008
39.	Furanolactone	Whole plant	Panchabhai et al., 2008
40.	Giloin	Whole plant	Panchabhai et al., 2008
41.	Giloinin	Whole plant	Panchabhai et al., 2008.
42.	Giloinsterol	Whole plant	Panchabhai et al., 2008
43.	Heptacosane	Leaf	Naik et al., 2014
44.	Heptacosanol	Whole plant	Panchabhai et al., 2008; Gupta et al., 2009
45.	Hexacosane	Leaf	Naik et al., 2014
46.	Hydroquinone	Leaf	Naik et al., 2014
47.	Hydroxyecdysone	Whole plant	Panchabhai et al., 2008
48.	Isobutyl phthalate	Leaf	Naik et al., 2014
49.	Isocolumbin	Whole plant	Panchabhai et al., 2008
50.	Jatrorrhizine	Whole plant	Sarma et al., 1998; Panchabhai et al., 2008; Singh and Chaudhuri, 2015
51.	Linalool	Leaf	Naik et al., 2014
52.	Magnoflorine	Whole plant	Sarma et al., 1998; Panchabhai et al., 2008; Singh and Chaudhuri, 2015

(*Continued*)

TABLE 1.2 (*Continued*) Major phytochemicals isolated and identified from different plant parts of *Tinospora cordifolia*

S. NO.	*COMPOUNDS*	*REPORTED PART*	*REFERENCES*
53.	Makisterone A	Whole plant	Panchabhai et al., 2008
54.	Myristic acid	Leaf	Naik et al., 2014
55.	*N*-hexacosanyl stearate	Stem	Gupta et al., 2009
56.	*N*-nonacosanol	Stem	Gupta et al., 2009
57.	*N*-octacosanol	stem	Gupta et al., 2009
58.	*N*-tetracontanol	Stem	Gupta et al., 2009
59.	*N*-triacontanyl palmitate	Stem	Gupta et al., 2009
60.	*N*-acetylasimilobine 2-O-B-D-glucopyranosyl-(1 → 2)-B-D-glucopyranoside (tinoscorside B)	Aerial part	Maurya, 1996; Maurya, 1998; Van Kiem et al., 2010
61.	*N*-formylasimilobine 2-O-B-D-glucopyranosyl-(1→ 2)-B-D-glucopyranoside (tinoscorside A)	Aerial part	Maurya, 1996; Maurya, 1998; Van Kiem et al., 2010
62.	Nonacosan-15-one	Whole plant	Panchabhai et al., 2008
63.	Nonanal	Leaf	Naik et al., 2014
64.	Octacosane	Leaf	Naik et al., 2014
65.	Palmatine	Stem	Sarma et al., 1998; Panchabhai et al., 2008; Singh and Chaudhuri, 2015
66.	Palmatoside C	Whole plant	Panchabhai et al., 2008
67.	Palmatoside P	Whole plant	Panchabhai et al., 2008
68.	Palmitic acid	Leaf	Naik et al., 2014
69.	Pentacosane	Leaf	Naik et al., 2014
70.	Pentadecanoic acid	Leaf	Naik et al., 2014
71.	Phenyl oxide	Leaf	Naik et al., 2014
72.	Phytol	Leaf	Naik et al., 2014
73.	Sinapyl 4-O-B-D-apiofuranosyl-(1 →6)-O-B-D-glucopyranoside (tinoscorside D)	Aerial part	Maurya, 1996; Maurya, 1998; Van Kiem et al., 2010

(*Continued*)

TABLE 1.2 (*Continued*) Major phytochemicals isolated and identified from different plant parts of *Tinospora cordifolia*

S. NO.	*COMPOUNDS*	*REPORTED PART*	*REFERENCES*
74.	Syringin	Whole plant	Panchabhai et al., 2008
75.	Syringinapiosylglycoside	Whole plant	Panchabhai et al., 2008
76.	Tembetarine	Whole plant	Panchabhai et al., 2008
77.	Tetrahydrofuran	Whole plant	Panchabhai et al., 2008
78.	Tetrahydropalmatine	Whole plant	Panchabhai et al., 2008
79.	Tincordin	Whole plant	Sivasubramanian et al., 2013
80.	Tinocordifolin	Whole plant	Panchabhai et al., 2008
81.	Tinocordifoliside	Whole plant	Panchabhai et al., 2008
82.	Tinoscorside C	Aerial part	Maurya, 1996; Maurya, 1998; Van Kiem et al., 2010
83.	Tinosporaclerodanoid	Stem	Ahmad et al., 2010
84.	Tinosporaclerodanol	Stem	Ahmad et al., 2010
85.	Tinosporafurandiol	Stem	Ahmad et al., 2010
86.	Tinosporafuranol	Stem	Ahmad et al., 2010
87.	Tinosporal	Whole plant	Panchabhai et al., 2008
88.	Tinosporaside	Stem	Pan et al., 2012
89.	Tinosporastery palmitate	Stem	Gupta et al., 2009
90.	Tinosporic acid	Whole plant	Panchabhai et al., 2008
91.	Tinosporide	Whole plant	Panchabhai et al., 2008
92.	Tinosporide	Whole plant	Sivasubramanian et al., 2013
93.	Tinosporidin	Whole plant	Panchabhai et al., 2008
94.	Tinosporin	Whole plant	Panchabhai et al., 2008
95.	Tinosporisides	Whole plant	Panchabhai et al., 2008
96.	Tinosporon	Whole plant	Panchabhai et al., 2008
97.	Tinosporon columbin	Whole plant	Panchabhai et al., 2008
98.	Triacosane	Leaf	Naik et al., 2014
99.	Yangambin	Stem	Bala et al., 2015a
100.	β-Sitosterol and δ-sitosterol	Stem	Sarma, 1998; Panchabhai et al., 2008; Ahmad et al., 2010; Singh and Chaudhuri, 2015.

This plant is still widely used as natural resource obtained from natural habitats. There are very few instances where it is also being cultivated for use/consumption. With very wide distribution and also the phenomenon of dioecy, the plant is expected to have very high genetic diversity, which largely remained unexplored. In recent years, genetic diversity has been studied by markers for random amplified polymorphic DNA, and inter simple sequence repeat primers (Rana et al., 2012; Singh et al., 2014; Paliwal et al., 2016; Gargi et al., 2017) have shown genetic variation within the population. Similarly, chemical diversity based on gender and location has been also explored to a limited extent by Bajpai et al. (2016). Recently, Lade et al. (2018) have published one of the most extensive studies on TC using 114 collections from different regions of India maintained in the germplasm garden in National Botanical Research Institute, Lucknow, India. They have observed a wide range of genetic diversity in the population, which may be due to the very wide distribution of this species and the phenomenon of dioecy. While understanding the chemo and genetic diversity is required for better utilization of this natural resource, it is even more important to initiate conservation strategies so that a balance in use and regeneration is maintained. With the broadening of the information on TC, the consumption of this drug is going to increase in years to come, and it will be a challenge to meet this demand with dwindling resources. It is therefore, very important to develop a strategy for understanding the diversity (bio and chemo), conservation, and judicious exploitation of this plant from natural vegetation besides efforts to grow/cultivate.

Phytochemical Analysis and Metabolite Diversity in *Tinospora cordifolia*

2

Plants produce a wide range of phytochemicals to attract pollinators, for self-defense from multitude of biotic and abiotic stresses, and to interact with their environments for survival and spread (Akula et al., 2011). Environment influences the qualitative and quantitative composition of phytochemicals in plants. Phytochemical screening of plant is important in order to identify the new bioactive compounds, which could become new leads or new drugs (Rout et al., 2009). Ever since its inception, mass spectrometry (MS) has been extensively used in the analysis of plant metabolites (Jorge et al., 2016). The advantage of MS techniques has been utilized in the characterization of the phytoconstituents of *Tinospora cordifolia* stem (TCS) using LC-MS/MS in recent years (Bajpai et al., 2016; Bajpai et al., 2017a, 2017b). Recent developments in the hyphenated liquid chromatography mass instrumentation (LC-MS) have led to quantum jump in the area of natural product chemistry. LC-MS has high selectivity and sensitivity for qualitative and quantitative analysis of bioactive compounds and their metabolites present in trace amount in plant extract (Reemtsma, 2003). With the considerable advancements in LC-MS and LC-MS/MS techniques, it is possible to obtain extremely precise chemical information such as UV chromatogram, exact/accurate mass, isotope distribution patterns, molecular formula, characteristic fragment pattern, structural elucidation, quantitation, and purity of a compound (De Villiers et al., 2016).

T. cordifolia (TC) is a dioecious plant and it is noteworthy that all the chemical analyses and biological activities carried out in the past somehow did not take this into account. In a recent study (Bajpai et al., 2016), stem samples from male and female plants were analyzed. This study also took into account the distribution and seasons to address edaphic factors and different growth phases regulated by seasons reflecting the metabolic status of the plants. The samples were obtained from different locations in India such as Lucknow, Jabalpur, and Kolkata during premonsoon (February to April) and postmonsoon (September to November) seasons. The plant material collected from wild is likely to have inherent chemical variations and/or an altogether different composition, which makes the quality management of this Ayurvedic drug (or any other plant drug) a challenge. Due to very wide distribution and extensive use of this plant in the traditional system of medicine, it is important to know the extent of variability, if any, in the metabolite profile due to geographical locations, seasons, and gender. Collection of plant material with the optimum concentration of bioactive compounds is of paramount importance for the preparation of effective herbal drugs. Therefore, it is important to determine the actual content(s) and variations in the bioactive compounds in the plant extract. Studies on this very important aspect however, are lacking. Recently, direct analysis in real time mass spectrometric (DART MS) and LC-MS/MS studies have made some headways in this direction. High performance liquid chromatography–electrospray ionization–quadrupole time of flight tandem mass spectrometry (HPLC-ESI-QTOF-MS/MS) and ultra performance liquid chromatography–electrospray ionization–quadrupole linear ion trap tandem mass spectrometry (UPLC-ESI-QqQ_{LIT}-MS/MS) were also applied for the phytochemical profiling and quantitate bioactive phytochemical in TCS.

2.1 DESCRIPTION OF PLANT MATERIALS

Male and female plants with leaf, stem, flowers, and fruits were collected from three sites of naturally growing population from the bank of river Gomati, Lucknow, for consecutive years from 2009, 2010, 2011 and 2012. Five male and five female plant samples (stem) were collected from flowering male and female plants. In all 18 collections were made from three major geographical locations from Jabalpur (JBL), Madhya Pradesh; Kolkata (KOL), West Bengal; and Lucknow (LKO), Uttar Pradesh (UP), in year 2009 in post monsoon (September to November) and 2010 in premonsoon (February to April) and post monsoon (September to November) seasons and followed in 2011 and 2012. Collected plant samples were identified and deposited in the departmental herbarium of Botany division of Central Drug Research Institute (CSIR-CDRI), Lucknow, India. Voucher specimen numbers with geographical location and collection period are reported in Table 2.1. Botanical reference standard (BRS) of *T. cordifolia* stem (TCS) was obtained from Tulsi Amrit Pvt Limited, Indore, India (Batch No. 10TC 1438).

2.2 REAGENT AND CHEMICALS

Ethanol (AR grade, Merck, Darmstadt, Germany) was used for phytochemical extraction. AR grade HCl, hexane, $CHCl_3$, ammonia, n-BuOH, and EtOAc (Sigma Aldrich, St. Louis, MO, USA) were used for alkaloidal treatment and fractionation of ethanolic extract. Acetonitrile, methanol (LC-MS grade, Sigma Aldrich, St. Louis, MO, USA) and formic acid (LC grade) were used for LC-MS studies. Ultrapure water was obtained from Direct-Q system (Millipore, Milford, MA, USA). The reference standards mangiferin, chrysin, and protocatechuic acid were from Extrasynthese (Genay, France). The reference standards of arginine, choline, isocorydine, tetrahydropalmatine, 20-hydroxyecdysone, orientin, isoorientin, luteolin, apigenin, eriodictyol, ferulic acid and quercetin were from Sigma Aldrich Ltd. (St. Louis, MO, USA). The purity of all the reference standards was ≥95%.

2.3 PREPARATION OF EXTRACTS

TCS samples were shade dried at room temperature and ground to fine powder. The powder (approximately 500 g) was suspended in a glass percolator with (1000 mL) ethanol sonicated for 30 min in ultrasonic water bath at 30°C for proper mixing and kept at room temperature for 24 h. The percolate was collected, and the extraction process was repeated for five times. The combined extract was filtered through filter paper (Whatman No. 1) and concentrated on Buchi rotary evaporator (Rotavapor-R2, Flawil, Switzerland) under reduced pressure of 20–50 kPa at 40°C resulting in a dark green semisolid mass (about 25–27 g, 5% yield).

TABLE 2.1 Geographical locations of *Tinospora cordifolia* collected and voucher specimen (VS)

S. NO.	*VS NO.*	*SAMPLE NAME*	*GEOGRAPHICAL LOCATIONS (INDIA)*	*PERIOD OF COLLECTION*
1.	DKM 24512	TCSL-POM-09	Lucknow (U.P.), 260 52′24.92″ N 800 52′26.74″ E, Elev 392 ft.	06-09-2009
2.	DKM 24520	TCSJ-POM-09	Jabalpur (M.P.), 230 06′44.10″ N 79057′05.51″ E, Elev 1296 ft.	12-11-2009
3.	DKM 24514	TCSK-POM-09	Kolkata (W.B.), 220 26′18.39″ N 880 23′57.60″ E, Elev 35 ft.	20-09-2009
4.	DKM 24528	TCSL-PRM-10	Lucknow (U.P.), 260 52′24.92″ N 800 52′26.74″ E, Elev 392 ft.	01-04-2010
5.	DKM 24526	TCSJ-PRM-10	Jabalpur (M.P.), 230 06′44.10″ N 79057′05.51″ E, Elev 1296 ft.	23-03-2010
6.	DKM 24522	TCSK-PRM-10	Kolkata (W.B.), 220 26′18.39″ N 880 23′57.60″ E, Elev 35 ft.	12-02-2010
7.	DKM 24537	TCSL-POM-10	Lucknow (U.P.), 260 52′24.92″ N 800 52′26.74″ E, Elev 392 ft.	10-09-2010
8.	DKM 24532	TCSJ-POM-10	Jabalpur (M.P.), 230 06′44.10″ N 79057′05.51″ E, Elev 1296 ft.	30-11-2010
9.	DKM 24529	TCSK-POM-10	Kolkata (W.B.), 220 26′18.39″ N 880 23′57.60″ E, Elev 35 ft.	15-09-2010
10.	DKM 24549	TCSL-PRM-11	Lucknow (U. P.), 260 52′24.92″ N 800 52′26.74″ E, Elev 392 ft.	25-04-2011
11.	DKM 24547	TCSJ-PRM-11	Jabalpur (M.P.), 230 06′44.10″ N 79057′05.51″ E, Elev 1296 ft.	17-04-2011
12.	DKM 24541	TCSK-PRM-11	Kolkata (W.B.), 220 26′18.39″ N 880 23′57.60″ E, Elev 35 ft.	04-03-2011
13.	KRA 24469	TCSL-POM-11	Lucknow (U.P.), 260 52′24.92″ N 800 52′26.74″ E, Elev 392 ft.	05-04-2011
14.	DKM 24559	TCSJ-POM-11	Jabalpur (M.P.), 230 06′44.10″ N 79057′05.51″ E, Elev 1296 ft.	14-11-2011
15.	DKM 24550	TCSK-POM-11	Kolkata (W.B.), 220 26′18.39″ N 880 23′57.60″ E, Elev 35 ft.	07-09-2011
16.	DKM 24570	TCSL-PRM-12	Lucknow (U.P.), 260 52′24.92″ N 800 52′26.74″ E, Elev 392 ft.	28-03-2012
17.	DKM 24574	TCSJ-PRM-12	Jabalpur (M.P.), 230 06′44.10″ N 79057′05.51″ E, Elev 1296 ft.	24-04-2012
18.	DKM 24566	TCSK-PRM-12	Kolkata (W.B.), 220 26′18.39″ N 880 23′57.60″ E, Elev 35 ft.	11-03-2012

TCS, *T. cordifolia* stem; J, Jabalpur (M.P., Madhya Pradesh); K, Kolkata (W.B., West Bengal); L, Lucknow (U.P., Uttar Pradesh); PRM, premonsoon; POM, postmonsoon; 09, 2009; 10, 2010; 11, 2011; 12, 2012.

Source: Reproduced from Bajpai et al., 2017a with permission from Elsevier.

The alkaloidal treatment and fractionation of TCS ethanolic extract was also performed to check the efficiency in detection of compounds. The ethanolic extract (dark green semi solid mass), was dissolved in 5% HCl and filtered. The filtrate was basified by liquid ammonia and extracted in order to get $CHCl_3$ and n-BuOH soluble alkaloid fractions. Water fraction was concentrated to yield water soluble alkaloids. The nonalkaloidal fraction was dissolved in water–methanol and partitioned with EtOAc and n-BuOH. The prepared fraction was analyzed by DART-MS.

A stock solution of each ethanolic extract (1.0 mg/mL) was prepared and diluted in methanol to final working concentrations and filtered through a 0.22 µm PVDF membrane MILLEX GV filter unit (Millex GV, PVDF, Merck Millipore, Darmstadt, Germany) and transferred into a HPLC autosampler vial prior to LC-MS analysis. Stock solutions were stored at –20°C for further use.

2.4 PREPARATION OF STANDARD AND STOCK SOLUTIONS

Stock solutions of reference standards containing choline, jatrorrhizine, magnoflorine, isocorydine, palmatine, tetrahydropalmatine, and 20-hydroxyecdysone were prepared in methanol (1 mg/mL). Stock solution was diluted with methanol to prepare 50 µg/mL solutions, which was injected into the HPLC-ESI-QTOF-MS system for authentication of these phytochemicals in TCS. The standard stock and working solutions were all stored at –20°C until use and vortexed prior to injection.

2.5 DART-TOF-MS ANALYSIS OF INTACT STEM SAMPLES

DART-TOF-MS measurements were recorded by using intact TCS as samples without any sample preparation or excessive processing. Stem pieces washed with Milli-Q water to clean the surface of any adhering particulate matter and air dried for 30 min in oven at 35°C and chopped in small pieces for DART-TOF-MS analysis. The mass spectrometer used was a JMS-100 TLC

(AccuTof) atmospheric pressure ionization TOF-MS (Jeol, Tokyo, Japan) fitted with a DART ion source. All samples were analyzed in 15 repeats to check the reproducibility of spectra from DART-TOF-MS analysis. Mass calibration was accomplished by including a mass spectrum of neat polyethylene glycol (PEG) (1:1 mixture PEG 200 and PEG 600) in the data file. The mass calibration was accurate to within ±0.002 u. Using the Mass Centre software, the elemental composition of selected peaks were determined. The DART-MS optimized analysis conditions for the phytochemical analysis of TC intact stem and ethanolic extract with alkaloidal fractions are given in the following section.

2.6 DART-TOF-MS PARAMETERS

Ion mode: Positive	Discharge electrode: 100 V
Resolution (FWHM): 6000	Grid electrode: 250 V
Orifice 1 potential: 28 V	Gas beam temperature: 300°C
Orifice 2 potential: 5 V	Sampling time: 20 s
Orifice 1 temperature: 100°C	Mass range: 50–1000 Da
Ring lens potential: 13 V	Peak voltage: 600 V
RF ion guide potential: 300 V	Detector voltage: –2300 V
Helium flow rate: 4.0 L/min	Acquisition rate: 3.0 s
Needle voltage: 3000 V	Sensitivity: High

FWHM, full width at half maximum

2.7 CHEMOMETRIC ANALYSIS TO STUDY DISCRIMINATION

The DART-MS data subjected to principal component analysis (PCA) to enable the determination of peaks defining the plant distribution (locations) and gender. PCA was performed with the STATISTICA software, windows version 7.0

(Stat Soft, Inc., USA). MS data for PCA analysis were extracted from DART mass spectrum of all the samples. Fifteen repeats of each plant from all localities and five repeats of herbal formulations in the mass range from *m/z* 100 to 800 were used for PCA. All ions having ≥5% peak intensity were selected for statistical analysis.

Factor analysis (FA) and PCA were used to reduce the data dimensionality. Groups were compared by Student's *t* test, one way and two way (analytes and sex) analysis of variance (ANOVA) followed by Tukey's post hoc test. The significance of mean difference within and in between the groups was done by Duncan multiple range test. Receiver operating characteristics (ROC) curve analysis was used to find out the discriminating ability and diagnostic accuracy of the metabolites (ions). Data were summarized as mean ± SD (standard deviation). PCA was applied to visualize the difference in abundance and discrimination of the gender. A two tailed $p < 0.05$ was considered statistically significant. By applying these tests, systematic errors can be significantly reduced, and it can be confirmed that only actual variations were applied to discriminate the samples.

2.8 LC-MS DATA PROCESSING FOR DISCRIMINATION

LC-MS analyses of samples were repeated three times to check the reproducibility of analysis. Although analysis was carried out in both positive and negative ion mode, positive ion mode was found more sensitive toward the metabolites and hence selected for analysis. The relative abundance of all *m/z* was calculated from the base peak ion chromatograms. The normalized abundance of all ions from all the samples was considered as variables, and each geographical location, season, and gender was considered as a class. It has been demonstrated earlier that the geographical locations, seasons, and gender can be classified even by using the relative ion abundance without exact knowledge of their concentrations.

2.9 DART-TOF-MS DATA PROCESSING FOR DISCRIMINATION

Fifteen spectra of each sample were recorded to check the reproducibility and the robustness of the optimized DART-TOF-MS method. The statistical model was tested and validated by randomly dividing the repeats of samples in known and unknown groups. Known samples were used to construct the model, and unknowns were used to test and validate the model. A total of 60% experimental results were used to build the PCA model for discriminating the gender and geographical sample of stem, and the remaining 40% experimental results were used for testing and validation of the PCA model so built.

2.10 VALIDATION STUDIES FOR CHEMOMETRIC METHODS

Validation of LC-ESI-MS method was performed using test data set. The validation consisted of studies on repeatability of results obtained from training data set, sensitivity, specificity, selectivity, and confidence interval (CI) of marker peaks. The chemometric method was first developed using training data set and then applied to test data set to check the validity. Experimental results of total 30 and 32 samples (training data set) were used to build the statistical model for geographical and seasonal variations, respectively, whereas experimental results of 12 and 10 samples (test data set) were used for validation and testing of the statistical model so built. In case of gender variation, total 12 experimental results (training data set) were used to build the statistical model, and 6 experimental results (test data set) were used to validate and test the model.

2.11 OPTIMIZATION OF DART-TOF-MS PARAMETERS

To optimize the DART-MS instrumental parameters, different potential, temperature, electrode and grid voltage, and gas flow rates were varied to determine the optimal ionization conditions for TCS to obtain maximum and accurate peaks (*m/z*) with no or minimum fragmentation. Finally, optimized condition as described earlier was accomplished to obtain characteristic fingerprint for TCS. The ethanolic (EtOH) extracts, alkaloidal fractions, and intact stem of TC were analyzed using DART-MS.

The alkaloidal extract was obtained by the method described previously based on an acid and base extraction process, leading to the production of the alkaloid rich fraction. The spectrum obtained from ethanolic extract was rich in peaks, and compound at *m/z* 342 was found most abundant. A representative DART mass spectrum of TC intact stems of gender (male and female) and geographical sample (West Bengal, Uttar Pradesh, and Madhya Pradesh) is given in Figure 2.1a and b.

2.12 DART-TOF-MS ANALYSIS OF MALE, FEMALE AND GEOGRAPHICAL SAMPLES

DART-MS commonly produces $[M+H]^+$ and $[M]^+$ molecular ion peaks, yielding relatively simple and clear mass spectra despite the complexity of plant matrix. Most of the compounds identified from intact TCS using DART-MS were protonated molecular ions. Compounds in TCS were identified by comparing their observed masses with the data bank, and their relative abundance was calculated by measuring the intensities of their peaks.

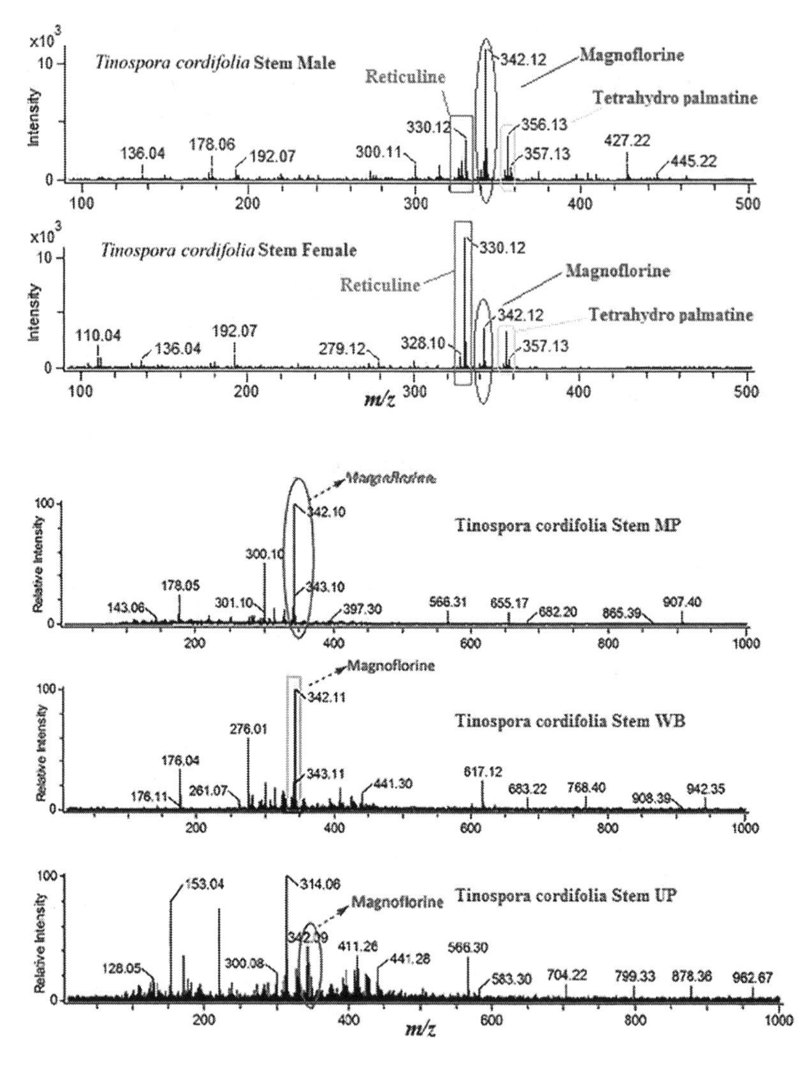

FIGURE 2.1 Gender representative DART mass spectra of *Tinospora cordifolia* male and female stem. Location specific, representative DART mass spectra of *Tinospora cordifolia* stem. (Reproduced from Bajpai et al., 2017a with permission from Elsevier.)

The confirmation of the identified compounds was made through high resolution mass measurement (HRMS) and their accurate molecular formula determinations. Peak at *m/z* 104.0763 $[M+H]^+$ could correspond to

N-methyl-2-pyrrolidone [C_5H_9NO], *m/z* 175.1172 [M+H]$^+$, arginine [$C_6H_{14}N_4O_2$], *m/z* 235.1687 [M+H]$^+$, 11-hydroxy mustakone [$C_{15}H_{22}O_2$], *m/z* 251.1628 [M+H]$^+$, tinocordifolin [$C_{15}H_{22}O_3$], *m/z* 294.1109 [M+H]$^+$, *N*-formylanonaine [$C_{18}H_{15}NO_3$], *m/z* 314.1381 [M+H]$^+$, *N*-trans-feruloyltyramine [$C_{18}H_{19}NO_4$], *m/z* 330.1691 [M+H]$^+$, [$C_{19}H_{23}NO_4$].

Alkaloids are the main compounds previously reported from the TCS; in the DART-MS, spectrum peak at *m/z* 338.1368 [M]$^+$ was found to correspond to jatrorrhizine [$C_{20}H_{20}NO_4^+$]; however, peak at *m/z* 342.1689 [M]$^+$ was found to correspond to magnoflorine [$C_{20}H_{24}NO_4^+$]; peak at *m/z* 352.1551 [M]$^+$ was found to correspond to palmatine [$C_{21}H_{22}NO_4^+$]; and peak at *m/z* 356.1869 [M]$^+$ was found to correspond to tetrahydropalmatine [$C_{21}H_{26}NO_4^+$]. Many other peaks tentatively identified were *m/z* 344.1476 [M+H]$^+$, *N*-trans-feruloyl-4-*O*-methyldopamine [$C_{19}H_{21}NO_5$]; *m/z* 359.1478 [M+H]$^+$, columbin [$C_{20}H_{22}O_6$]; *m/z* 397.2211 [M+H]$^+$, tinocordiside [$C_{21}H_{32}O_7$]; *m/z* 413.2148 [M+H]$^+$, tinocordifolioside [$C_{21}H_{32}O_8$]; and *m/z* 415.3927 [M+H]$^+$, β-sitosterol [$C_{29}H_{50}O_1$]. Several other peaks were unidentified due to lack of sufficient data and literature support. The exact mass values measured in TCS using the DART-TOF-MS are given in Table 2.2. So far, peaks observed at *m/z* 100, 175, 235, 251, 294, 314, 330, 338, 342, 344, 352, 356, 359, 357, 413, and 415 were *N*-methyl-2-pyrrolidone, arginine, 11-hydroxymustakone, tinocordifolin, *N*-formylanonaine, *N*-trans-feruloyltyramine, reticuline, jatrorrhizine, magnoflorine, *N*-trans-feruloyl-4-*O*-methyldopamine, palmatine, tetrahydropalmatine, columbin, tinocordiside, tinocordifolioside, and β-sitosterol, respectively.

Male and female intact stem from different geographical locations were recorded by DART-MS, and spectra obtained were reproducible. All the ions with a relative intensity above 3% were taken and compared on the basis of % ionization.

Fifteen repeats of each sample were carried out, and the averaged result was utilized for the analysis. Comparative fingerprint spectra of the stem samples are shown in Figure 2.1. Sixteen compounds were identified in all the samples based on their exact mass, molecular formula, and literature reports. The spectra showed characteristic fingerprint patterns for each stem sample. The DART-MS spectra revealed the variations in the distribution of some of the most common peaks according to abundance and presence. It is clear from the spectra that relative abundance of *m/z* 330, 342, and 356 corresponding to reticuline, magnoflorine, and tetrahydropalmatine respectively, is different in male and female stem. Thus, the DART-MS fingerprints would be useful in selecting the right population for medicinal purposes on the basis of the relative abundance of the bioactive compounds.

TABLE 2.2 Exact mass data from the direct analysis in real time mass spectra of *Tinospora cordifolia* stem

	MASS					*PEAK INTENSITY IN % (N = 15)*	
S. NO.	*THEORETICAL M/Z*	*OBSERVED M/Z*	*Δ ERROR (PPM)*	*FORMULA*	*IDENTIFICATION*	*INTACT STEM*	*ETOH EXTRACT OF STEM*
1.	100.0775 $[M+H]^+$	100.0763	1.2	C_5H_9NO	*N*-methyl-2-pyrrolidone	2.5	0.6
2.	175.1190 $[M+H]^+$	175.1172	1.7	$C_6H_{14}N_4O_2$	Arginine	3.3	0.8
3.	235.1693 $[M+H]^+$	235.1687	0.6	$C_{15}H_{22}O_2$	11-Hydroxymustakone	4.2	Bdl
4.	251.1642 $[M+H]^+$	251.1628	1.4	$C_{15}H_{22}O_3$	Tinocordifolin	2.0	5.6
5.	294.1125 $[M+H]^+$	294.1109	1.6	$C_{18}H_{15}NO_3$	*N*-Formylanonaine	3.1	23.2
6.	314.1387 $[M+H]^+$	314.1381	0.6	$C_{18}H_{19}NO_4$	*N*-trans-feruloyltyramine	6.2	28.5
7.	330.1700 $[M+H]^+$	330.1691	0.8	$C_{19}H_{23}NO_4$	Reticuline	32.4	12.0
8.	338.1387 $[M]^+$	338.1368	1.9	$C_{20}H_{20}NO_4^+$	Jatrorrhizine	11.6	16.5
9.	342.1700 $[M]^+$	342.1689	1.64	$C_{20}H_{24}NO_4^+$	Magnoflorine	100.0	100.0
10.	344.1492 $[M+H]^+$	344.1476	1.6	$C_{19}H_{21}NO_5$	*N*-trans-feruloyl-4-*O*-methyldopamine	28.5	17.3
11.	352.1543 $[M]^+$	352.1551	–0.8	$C_{21}H_{22}NO_4^+$	Palmatine	8.2	6.8
12.	356.1856 $[M]^+$	356.1869	–1.3	$C_{21}H_{26}NO_4^+$	Tetrahydropalmatine	37.6	16.5
13.	359.1489 $[M+H]^+$	359.1478	1.1	$C_{20}H_{22}O_6$	Columbin	2.7	Bdl
14.	397.2220 $[M+H]^+$	397.2211	0.9	$C_{21}H_{32}O_7$	Tinocordiside	3.8	31.7
15.	413.2170 $[M+H]^+$	413.2148	2.1	$C_{21}H_{32}O_8$	Tinocordifolioside	4.3	Bdl
16.	415.3934 $[M+H]^+$	415.3927	0.7	$C_{29}H_{50}O_1$	β-Sitosterol	6.1	Bdl

Bdl, below detection level.
Source: Reproduced from Bajpai et al., 2017a with permission from Elsevier.

2.13 PRINCIPAL COMPONENT ANALYSIS USING DART-TOF-MS FINGERPRINTS

PCA was used for processing of DART-MS derived fingerprints for stem samples of TC from different geographical location and gender. The stem PCA showed the differences in chemical profile of male and female. In the spectra of male and female stem, total 104 peaks were extracted (intensity ≥ 3.0) from *m/z* 100–750 Da range, which were treated as variables in the PCA. PCA was initially run based on 104 variables of male and female samples and 173 variables of geographical location samples. The first two principal components PC1 and PC2, respectively, hold 37.37% and 26.50% of the total variability in the male and female stem samples. In the case of geographical locations, PC1 and PC2 hold 71.37% and 23.96%, respectively. Thus, PCs were able to explain about 63.87% variability in male and female plants and about 95.33% of the total variability in the data. Out of 104 peaks, only 9 variables (peaks) at *m/z* 110, 313, 315, 330, 342, 375, 397, 463, and 575 were required to explain the total 63.87% variability (Figure 2.2), which brings the clear discrimination based on the gender.

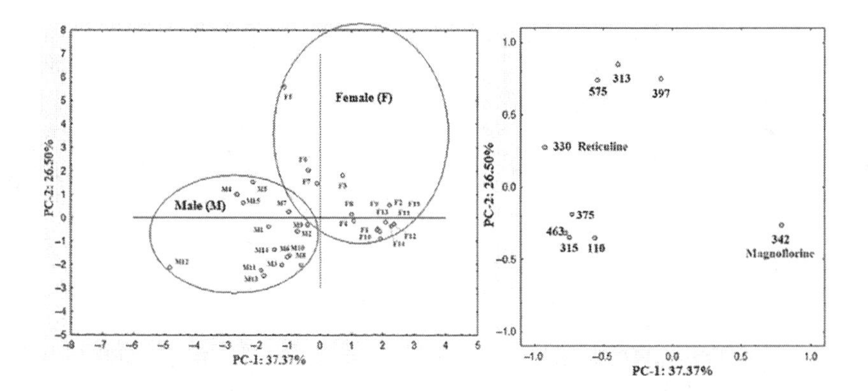

FIGURE 2.2 Validated principal component analysis (PCA) score plot and loading plot showing discrimination between male and female *Tinospora cordifolia* stem. (Reproduced from Bajpai et al., 2017a with permission from Elsevier.)

In case of geographical locations, the stem samples from Lucknow and Jabalpur were very similar, whereas West Bengal sample was entirely different. However, out of 173 peaks, only 16 variables (peaks) at *m/z* 153, 176, 219, 282, 295, 300, 308, 314, 330, 342, 344, 359, 427, 429, 445, and 566 were able to explain the total 95.33% variability, which differentiated the geographical locations of the TC samples. Therefore, all the 9 and 16 marker peaks were responsible for gender and geographical discrimination in the samples.

A high throughput method for differentiation of male and female plants and different geographical location samples of TCS using DART-MS fingerprints and chemometric analysis was developed. It is considered that this approach will be helpful in the selection of the best plant population.

2.14 HPLC-ESI-QTOF-MS/MS CONDITIONS

All the TCS samples were analyzed on Agilent 6520 QTOF-MS connected with Agilent 1200 HPLC system via Dual ESI interface (Agilent Technologies, USA). The 1200 HPLC system consisted of a quaternary pump (G1311A), online vacuum degasser (G1322A), autosampler (G1329A), and a diode array detector (G1315D). The HPLC separation was carried out on a Thermo Betasil C_8 column (250 mm × 4.5 mm, 5 μ) equipped with Thermo Betasil C_8 HPLC guard cartridge (10 mm × 4.0 mm, 5 μ) operated at 25°C. The mobile phase consisted of 0.1% formic acid aqueous solution (A) and acetonitrile (B) with a flow rate of 0.5 mL/min under the gradient program of 30–40% (B) for initial 5 min and subsequently, 40–55% (B) from 5 to 10 min, 55–60% (B) from 10 to 18 min, 60–75% (B) from 18 to 28 min, 75–90% (B) from 28 to 38 min, 90–92% (B) from 38 to 48 min, 92-40% (B) from 48 to 60 min followed by initial 30% (B) in 60–65 min. The sample injection volume was 3 μL. The diode array detector recorded UV spectra in the range 190–400 nm and was set to monitor at 254 and 280 nm.

MS analysis was performed on Agilent 6520 QTOF-MS in positive ESI mode. The resolving power of QTOF analyzer was set above 10000 (full width at half maximum), and spectra were acquired within a mass range of *m/z* 50–1,000 Da using nitrogen as nebulizing, drying, and collision gas. Capillary temperature was set to 350°C, nebulizer pressure to 40 psi and the drying gas flow rate to 10 L/min. Ion source parameters such as Vcap, fragmentor, skimmer, and octapole RF peak voltages were set to 3,500, 150, 65, and 750 V, respectively. Collision energy values for MS/MS experiments were fixed at 25, 30, and 35 eV for all the selected masses. Accurate mass measurements were carried out by the auto mass calibration method, which was performed using the external mass calibration solution (ESI-L Low Concentration Tuning Mix; Agilent calibration solution B). The chromatographic and MS analysis and prediction of chemical formula including the exact mass calculation were performed by Mass Hunter software version B.04.00 build 4.0.479.0 (Agilent Technology).

2.15 SCREENING OF PHYTOCHEMICALS BY HPLC-ESI-QTOF-MS/MS ANALYSIS

Thirty six compounds were identified and characterized using HPLC-ESI-QTOF-MS/MS in extracts. The base peak chromatogram of TCS is shown in Figure 2.3. The compounds were designated by the peak numbers. The retention time, molecular ions $[M]^{+\cdot}$ or protonated molecular ions $[M+H]^+$, calculated and observed *m/z* values, molecular formulae, error values, and MS/MS data of detected metabolites are shown in Table 2.3.

2.16 GUANIDINO COMPOUNDS

Many naturally occurring guanidino compounds were identified in natural products with a strong basic guanidine moiety. Peak 8 was positively identified by comparison with reference standard as arginine (*m/z* 175.1184, $C_6H_{14}N_4O_2$). Peaks 1 and 4 produced the most prominent ion at *m/z* 70.0651. Neutral guanidine (59 Da) loss at *m/z* 73.0647 and 71.0653 was also obtained for 1 and 4, respectively. Peaks 1 and 4 were tentatively identified as γ-guanidino butyl alcohol and γ-guanidinobutyraldehyde, respectively.

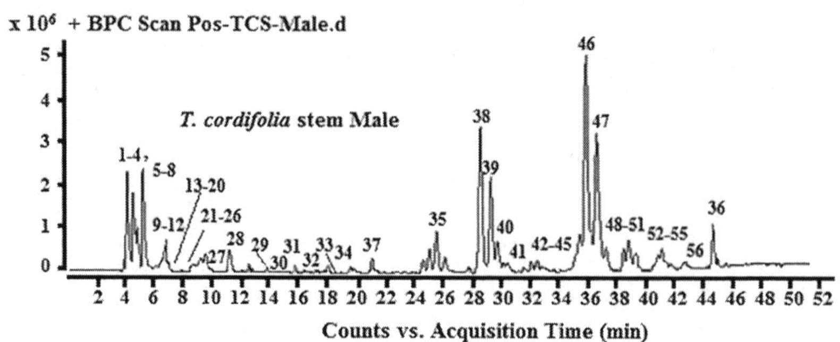

FIGURE 2.3 Base peak chromatogram of *Tinospora cordifolia* stem (TCS) extract. (Reproduced from Bajpai et al., 2017a with permission from John Wiley and Sons.)

Similarly, peak 3 showed the ions at *m/z* 87.0441 due to loss of neutral guanidine (59 Da) and *m/z* 60.0569 due to protonated guanidine ion $[CH_5N_3+H]^+$ and identified as γ-guanidino butyric acid (Peer and Sharma, 1989). Collision-induced dissociation of acyclic guanidines showed a characteristic loss of neutral guanidine (Peer and Sharma, 1989). Therefore, characteristic ions for guanidino compounds (1, 3, 4, and 8) were identified as $[M+H-CH_5N_3]^+$ and $[CH_5N_3+H]^+$ ions.

2.17 PROTOBERBERINE ALKALOIDS

Quaternary protoberberine alkaloids were easily ionized and gave abundant $[M]^+$ ions in their positive ESI-MS spectra due to the presence of quaternary nitrogen.

TABLE 2.3 Compounds identified in male and female *Tinospora cordifolia* stem based on mass spectrometric analysis

		MASS						*PEAK AREA (N = 3)*	
PEAK NO.	*RT (MIN)*	*CALCULATED M/Z*	*OBSERVED M/Z*	*Δ ERROR (PPM)*	*FORMULA*	*FRAGMENTS M/Z (ABUNDANCE)*	*IDENTIFICATION*	*FEMALE*	*MALE*
1.	4.26	132.1131 $[M+H]^+$	132.1130	0.76	$C_5H_{13}N_3O$	115.0840 (5), 90.0912 (18), 73.0647 (46), 70.0651 (100), 60.0567 (39)	γ-Guanidino butyl alcohol[a]	3.5E+05	9.6E+05
2.	4.33	104.1070 $[M]^+$	104.1069	0.96	$C_5H_{14}NO^+$	60.0812 (100), 58.0655 (92), 45.0342 (56), 44.0503 (34), 43.0187 (11)	Choline[e‡]	1.5E+06	2.4E+06
3.	4.35	146.0924 $[M+H]^+$	146.0934	–6.84	$C_5H_{11}N_3O_2$	129.0011 (1), 111.0535 (10), 87.0440 (90), 87.0441 (100), 86.063 (68), 69.0439 (24), 60.569 (17), 45.0346 (43)	γ-Guanidino butyric acid[a]	5.7E+04	4.4E+04

(*Continued*)

TABLE 2.3 (*Continued*) Compounds identified in male and female *Tinospora cordifolia* stem based on mass spectrometric analysis

		MASS						*PEAK AREA (N = 3)*	
PEAK NO.	*RT (MIN)*	*CALCULATED M/Z*	*OBSERVED M/Z*	*Δ ERROR (PPM)*	*FORMULA*	*FRAGMENTS M/Z (ABUNDANCE)*	*IDENTIFICATION*	*FEMALE*	*MALE*
4.	4.37	130.0975 $[M+H]^+$	130.0967	6.15	$C_5H_{11}N_3O$	113.0915 (1), 112.0860 (2), 90.0903 (13), 71.0653 (6), 70.651 (100), 60.0571 (4)	γ-Guanidino butyraldehyde[a]	2.9E+04	1.8E+06
5.	4.67	118.0863 $[M]^+$	118.0859	3.39	$C_5H_{12}NO_2^+$	118.0859 (100), 59.0734 (56)	Glycine betaine[e]	5.4E+05	4.6E+05
6.	4.75	144.1019 $[M]^+$	144.1016	2.08	$C_7H_{14}NO_2^+$	144.1018(100), 84.0823(10), 58.0672(10)	Proline betaine[e]	2.8E+04	4.8E+04

(*Continued*)

TABLE 2.3 (*Continued*) Compounds identified in male and female *Tinospora cordifolia* stem based on mass spectrometric analysis

		MASS						*PEAK AREA (N = 3)*	
PEAK NO.	*RT (MIN)*	*CALCULATED M/Z*	*OBSERVED M/Z*	*Δ ERROR (PPM)*	*FORMULA*	*FRAGMENTS M/Z (ABUNDANCE)*	*IDENTIFICATION*	*FEMALE*	*MALE*
7.	4.94	272.1281 $[M+H]^+$	272.1271	3.67	$C_{16}H_{17}NO_3$	255.0926(20), 237.0890(40), 164.0678 (2). 161.0591(30), 143.0712 (30), 115.0583(37), 107.0503(100), 77.0449(18)	Norcoclaurine[c]	1.9E+05	1.2E+05
8.	4.95	175.1190 $[M+H]^+$	175.1184	3.43	$C_6H_{14}N_4O_2$	175.1191(12), 158.0938(12), 130.0977(75), 116.0693 (15), 70.0603 (100), 60.0569 (9)	Arginine[a‡]	1.4E+06	3.5E+05
9.	5.02	342.1700 $[M+H]^+$	342.1705	−1.46	$C_{20}H_{23}NO_4$	192.0992, 178.0865 (100), 151.0740 (7), 137.0607 (5), 123.0817	Tetrahydrojatror-rhizine[d]	8.0E+05	8.8E+05

(*Continued*)

TABLE 2.3 (*Continued*) Compounds identified in male and female *Tinospora cordifolia* stem based on mass spectrometric analysis

		MASS						*PEAK AREA (N = 3)*	
PEAK NO.	*RT (MIN)*	*CALCULATED M/Z*	*OBSERVED M/Z*	*Δ ERROR (PPM)*	*FORMULA*	*FRAGMENTS M/Z (ABUNDANCE)*	*IDENTIFICATION*	*FEMALE*	*MALE*
10.	5.4	330.1700 $[M+H]^+$	330.1693	2.12	$C_{19}H_{23}NO_4$	299.1273 (4.59), 267.1002 (4.18), 192.1019 (100), 175.0752 (22.35), 151.0753 (6.03), 143.0491 (19.76), 122.0359 (1), 137.057 (40.04), 115.0542 (4.12), 44.0504 (2)	Reticuline[c]	8.0E+05	3.1E+05

(*Continued*)

TABLE 2.3 (*Continued*) Compounds identified in male and female *Tinospora cordifolia* stem based on mass spectrometric analysis

		MASS						*PEAK AREA (N = 3)*	
PEAK NO.	*RT (MIN)*	*CALCULATED M/Z*	*OBSERVED M/Z*	*Δ ERROR (PPM)*	*FORMULA*	*FRAGMENTS M/Z (ABUNDANCE)*	*IDENTIFICATION*	*FEMALE*	*MALE*
11.	5.56	342.1700 $[M]^+$	342.1689	3.21	$C_{20}H_{24}NO_4^+$	297.1123 (58.78), 282.0881 (23.75), 265.0860 (71.02), 237.0909 (19.75), 219.0815 (7.48), 191.0861 (5.11), 58.0664(100)	Laurifoline[b]	2.5E+05	3.1E+05

(*Continued*)

TABLE 2.3 (*Continued*) Compounds identified in male and female *Tinospora cordifolia* stem based on mass spectrometric analysis

		MASS						*PEAK AREA* (N = *3)*	
PEAK NO.	*RT (MIN)*	*CALCULATED M/Z*	*OBSERVED M/Z*	*Δ ERROR (PPM)*	*FORMULA*	*FRAGMENTS M/Z (ABUNDANCE)*	*IDENTIFICATION*	*FEMALE*	*MALE*
12.	5.58	314.1751 $[M]^+$	314.1752	−0.32	$C_{19}H_{24}NO_3^+$	269.1176 (11.12), 237.0906 (9.7), 209.0962 (8.15), 207.1207 (1), 192.1012 (6.9), 175.0763 (8.7), 145.062 (10), 143.0496 (19.78), 121..652 (9.5), 115.0544 (10.92), 107.0494 (100), 58.0663 (64.03)	Oblongine[c]	1.7E+05	1.0E+06

(*Continued*)

TABLE 2.3 (*Continued*) Compounds identified in male and female *Tinospora cordifolia* stem based on mass spectrometric analysis

		MASS						*PEAK AREA (N = 3)*	
PEAK NO.	*RT (MIN)*	*CALCULATED M/Z*	*OBSERVED M/Z*	*Δ ERROR (PPM)*	*FORMULA*	*FRAGMENTS M/Z (ABUNDANCE)*	*IDENTIFICATION*	*FEMALE*	*MALE*
13.	7.03	342.1700 $[M]^+$	342.1689	3.21	$C_{20}H_{24}NO_4^+$	297.1123 (58.78), 282.0881 (23.75), 265.0860 (71.02), 237.0909 (19.75), 219.0815 (7.48), 191.0861 (5.11), 58.0664(100)	Magnoflorine[b‡]	1.8E+06	2.5E+06
14.	7.06	481.3160 $[M+H]^+$	481.3164	−0.83	$C_{27}H_{44}O_7$	463.3088 (16), 445.2938 (100), 427.2832 (31), 409.2730 (10), 371.2211 (65), 303.1952 (10), 301.1801 (6), 165.1274 (30), 125.0958 (6)	20-Hydroxyecdysone[f‡]	3.1E+05	4.0E+05

(*Continued*)

TABLE 2.3 (*Continued*) Compounds identified in male and female *Tinospora cordifolia* stem based on mass spectrometric analysis

		MASS						PEAK AREA (N = 3)	
PEAK NO.	RT (MIN)	CALCULATED M/Z	OBSERVED M/Z	Δ ERROR (PPM)	FORMULA	FRAGMENTS M/Z (ABUNDANCE)	IDENTIFICATION	FEMALE	MALE
15.	7.19	342.1700 $[M+H]^+$	342.1705	−1.46	$C_{20}H_{23}NO_4$	311.1272 (10), 296.1036 (70), 279.1008 (100), 264.0774 (35), 248.0774 (15), 236.0834 (12), 219.0823 (5), 208.0863 (5), 58.065 (4)	Isocorydine[b‡]	8.8E+04	1.1E+05
16.	7.3	324.1230 $[M]^+$	324.1216	4.32	$C_{19}H_{18}NO_4^+$	309.0978 (100), 294.0747 (47), 281.1051 (2), 266.0802 (12), 238.0803 (2)	Stepharanine[d]	4.4E+04	7.9E+04

(*Continued*)

TABLE 2.3 (*Continued*) Compounds identified in male and female *Tinospora cordifolia* stem based on mass spectrometric analysis

		MASS						*PEAK AREA* (N = 3)	
PEAK NO.	*RT (MIN)*	*CALCULATED M/Z*	*OBSERVED M/Z*	*Δ ERROR (PPM)*	*FORMULA*	*FRAGMENTS M/Z (ABUNDANCE)*	*IDENTIFICATION*	*FEMALE*	*MALE*
17.	7.47	356.1856 $[M]^+$	356.1841	4.21	$C_{21}H_{26}NO_4^+$	311.1264 (8.62), 296.1027 (19.41), 279.1010 (53.13), 264.0789 (27.02), 248.0836 (19.03), 236.0831 (20.74), 220.0833 (4.16), 208.0907 (3.23), 58.0663 (100)	Menisperine[b]	2.3E+03	2.0E+03

(*Continued*)

TABLE 2.3 (*Continued*) Compounds identified in male and female *Tinospora cordifolia* stem based on mass spectrometric analysis

		MASS						*PEAK AREA (N = 3)*	
PEAK NO.	*RT (MIN)*	*CALCULATED M/Z*	*OBSERVED M/Z*	*Δ ERROR (PPM)*	*FORMULA*	*FRAGMENTS M/Z (ABUNDANCE)*	*IDENTIFICATION*	*FEMALE*	*MALE*
18.	7.62	413.2170 $[M+H]^+$	413.2155	3.63	$C_{21}H_{32}O_8$	251.1623 (42), 233.1538 (25), 215.1374 (14), 175.1105 (100), 159.0708 (6)	Tinocordifolioside[g]	5.4E+04	7.7E+04
19.	7.66	356.1856 $[M]^+$	356.1842	3.93	$C_{21}H_{26}NO_4^+$	356.1865(6), 311.1273(40), 296.1041(60), 281.0831(16), 280.1116(100), 265.0868(13)	Xanthoplanine[b]	2.8E+03	2.2E+03

(*Continued*)

TABLE 2.3 (*Continued*) Compounds identified in male and female *Tinospora cordifolia* stem based on mass spectrometric analysis

		MASS						*PEAK AREA (N = 3)*	
PEAK NO.	*RT (MIN)*	*CALCULATED M/Z*	*OBSERVED M/Z*	*Δ ERROR (PPM)*	*FORMULA*	*FRAGMENTS M/Z (ABUNDANCE)*	*IDENTIFICATION*	*FEMALE*	*MALE*
20.	7.7	495.3316 $[M+H]^+$	495.3308	1.62	$C_{28}H_{46}O_7$	477.3148 (13), 459.3087 (100), 441.2965 (26), 423.2837 (6), 371.2186 (63), 357.2043 (39), 339.1909 (9), 329.2060 (12), 303.1937 (12), 219.1357 (13), 179.1421 (23), 139.1098 (6), 113.0971 (8)	Makisterone A[f]	1.5E+04	8.7E+03
21.	8.27	374.1598 $[M+H]^+$	374.1573	6.68	$C_{20}H_{23}NO_6$	329.1007 (62), 314.0776 (23), 269.0808 (15) 255.0655 (100), 240.0358 (5), 223.0397 (11), 137.0601 (6), 107.0846 (5), 58.0655 (11)	Chiloenamine[b]	1.0E+05	1.4E+05

(*Continued*)

TABLE 2.3 (*Continued*) Compounds identified in male and female *Tinospora cordifolia* stem based on mass spectrometric analysis

		MASS						*PEAK AREA (N = 3)*	
PEAK NO.	*RT (MIN)*	*CALCULATED M/Z*	*OBSERVED M/Z*	*Δ ERROR (PPM)*	*FORMULA*	*FRAGMENTS M/Z (ABUNDANCE)*	*IDENTIFICATION*	*FEMALE*	*MALE*
22.	8.78	356.1856 $[M]^+$	356.1841	4.21	$C_{21}H_{26}NO_4^+$	356.1870(5), 192.1018(100), 177.0778(10)	*N*-methyl tetrahydro-columbamine[d]	3.4E+05	4.1E+05
23.	8.78	356.1856 $[M]^+$	356.1841	4.21	$C_{21}H_{26}NO_4^+$	340.1546 (4), 192.1016 (100), 190.087(3), 165.0906 (25), 150.0661 (6)	Tetrahydropalmatine[d‡]	5.9E+05	4.6E+05
24.	9.6	481.316 $[M+H]^+$	481.3164	–0.83	$C_{27}H_{44}O_7$	463.3088 (16), 445.2938 (100), 427.2832 (31), 409.2730 (10), 371.2211 (65), 303.1952 (10), 301.1801 (6), 165.1268 (30), 125.0958 (6)	Ecdysterone[f]	1.3E+04	1.1E+04

(*Continued*)

TABLE 2.3 (*Continued*) Compounds identified in male and female *Tinospora cordifolia* stem based on mass spectrometric analysis

		MASS						*PEAK AREA (N = 3)*	
PEAK NO.	*RT (MIN)*	*CALCULATED M/Z*	*OBSERVED M/Z*	*Δ ERROR (PPM)*	*FORMULA*	*FRAGMENTS M/Z (ABUNDANCE)*	*IDENTIFICATION*	*FEMALE*	*MALE*
25.	9.64	338.1387 $[M]^+$	338.1379	2.37	$C_{20}H_{20}NO_4^+$	323.1130 (100), 322.1073 (99), 308.0910 (42.01), 306.0761 (30.5), 294.1131 (63.05), 279.089 (20.1), 280.0985 (23.32), 265.0739 (10.57)	Columbamine[d]	2.4E+05	2.2E+05
26.	10.74	338.1387 $[M]^+$	338.1387	0.06	$C_{20}H_{20}NO_4^+$	323.1130 (92), 322.1073 (100), 308.0910 (42.01), 306.0761 (30.5), 294.1131 (63.05), 279.089 (20.1), 280.0985 (23.32), 265.0739 (10.57)	Jatrorrhizine[d‡]	1.8E+05	3.5E+05

(*Continued*)

TABLE 2.3 (*Continued*) Compounds identified in male and female *Tinospora cordifolia* stem based on mass spectrometric analysis

		MASS						*PEAK AREA (N = 3)*	
PEAK NO.	*RT (MIN)*	*CALCULATED M/Z*	*OBSERVED M/Z*	*Δ ERROR (PPM)*	*FORMULA*	*FRAGMENTS M/Z (ABUNDANCE)*	*IDENTIFICATION*	*FEMALE*	*MALE*
27.	11.4	251.1642 $[M+H]^+$	251.1631	4.38	$C_{15}H_{22}O_3$	233.1532 (5), 215.1420 (8), 205.1572 (2), 197.1300 (3), 187.1420 (6), 175.1109 (100), 159.0797 (13), 145.1002 (11), 135.0798 (11), 121.0642 (11), 109.0645 (80, 93.0698 (5), 81.0708 (3)	Tinocordifolin[9]	8.7E+02	1.7E+03

(*Continued*)

TABLE 2.3 (*Continued*) Compounds identified in male and female *Tinospora cordifolia* stem based on mass spectrometric analysis

		MASS						PEAK AREA (N = 3)	
PEAK NO.	RT (MIN)	CALCULATED M/Z	OBSERVED M/Z	Δ ERROR (PPM)	FORMULA	FRAGMENTS M/Z (ABUNDANCE)	IDENTIFICATION	FEMALE	MALE
28.	11.62	352.1543 $[M]^+$	352.1546	–0.85	$C_{21}H_{22}NO_4^+$	337.1280 (80), 336.1223 (100), 322.1071 (37.91), 320.0917 (20.9), 308.1275 (70.33), 294.1111 (20.95)	Palmatine[d‡]	5.0E+05	6.4E+05
29.	13.9	344.1492 $[M+H]^+$	344.1492	0.00	$C_{19}H_{21}NO_5$	177.0537 (100), 151.0741 (14), 149.0580 (9), 145.0268 (90), 137.0603 (7), 117.0328 (16)	*N*-trans-feruloyl-4-*O*-methyldopamine[e]	4.7E+04	1.7E+04

(*Continued*)

TABLE 2.3 (*Continued*) Compounds identified in male and female *Tinospora cordifolia* stem based on mass spectrometric analysis

		MASS						*PEAK AREA* (N = 3)	
PEAK NO.	*RT (MIN)*	*CALCULATED M/Z*	*OBSERVED M/Z*	*Δ ERROR (PPM)*	*FORMULA*	*FRAGMENTS M/Z (ABUNDANCE)*	*IDENTIFICATION*	*FEMALE*	*MALE*
30.	14.1	314.1387 $[M+H]^+$	314.1378	2.86	$C_{18}H_{19}NO_4$	177.0538 (98), 149.0856 (4), 145.0276 (100), 121.0642 (92), 117.0331 (14), 93.0698 (13), 89.0391 (1), 77.0388 (5)	*N*-trans-feruloyltyramine[e]	5.7E+05	1.8E+05
31.	16.12	328.1543 $[M+H]^+$	328.1539	1.22	$C_{19}H_{21}NO_4$	297.1102 (15), 282.0902 (12), 267.0827 (11), 265.0856 (100), 264.050 (2), 254.0754 (3), 250.0752 (7), 237.0891(28), 233.0645 (26), 222.0646 (6), 205.0647 (21)	Isoboldine[b]	1.4E+03	1.2E+03

(*Continued*)

TABLE 2.3 (*Continued*) Compounds identified in male and female *Tinospora cordifolia* stem based on mass spectrometric analysis

		MASS						*PEAK AREA (N = 3)*	
PEAK NO.	*RT (MIN)*	*CALCULATED M/Z*	*OBSERVED M/Z*	*Δ ERROR (PPM)*	*FORMULA*	*FRAGMENTS M/Z (ABUNDANCE)*	*IDENTIFICATION*	*FEMALE*	*MALE*
32.	17.12	235.1693 $[M+H]^+$	235.1692	0.43	$C_{15}H_{22}O_2$	217.1629 (9), 189.16596 (6), 175.1129 (20), 161.0956 (100), 157.0995 (10), 147.1143 (4), 147.0803 (16), 135.0799 (36), 133.1020 (19), 121.0649 (30), 119.0847 (5), 109.1005 (12), 105.0722 (8), 93.0685 (8), 83.0486 (5), 69.0696 (8)	11-hydroxymustakone[g]	9.0E+03	5.5E+03

(*Continued*)

TABLE 2.3 (*Continued*) Compounds identified in male and female *Tinospora cordifolia* stem based on mass spectrometric analysis

		MASS						*PEAK AREA (N = 3)*	
PEAK NO.	*RT (MIN)*	*CALCULATED M/Z*	*OBSERVED M/Z*	*Δ ERROR (PPM)*	*FORMULA*	*FRAGMENTS M/Z (ABUNDANCE)*	*IDENTIFICATION*	*FEMALE*	*MALE*
33.	18.81	328.1543 $[M+H]^+$	328.1539	1.22	$C_{19}H_{21}NO_4$	297.1149 (7), 282.09029(15), 267.0821 (6), 265.0854 (70), 264.053 (5), 254.0748 (5), 250.0615 (31), 237.0882(78), 233.0584(100), 222.0674 (61), 205.0641 (90)	Boldine[b]	5.7E+03	3.9E+03
34.	19.55	354.1336 $[M+H]^+$	354.1316	5.65	$C_{20}H_{19}NO_5$	339.1101 (100), 310.1074, 324.0866, 321.0982 (52.23), 306.0746 (17.92), 296.0904 (29.92)	8-Oxojatrorrhizine[d]	6.4E+02	5.1E+02

(*Continued*)

TABLE 2.3 (*Continued*) Compounds identified in male and female *Tinospora cordifolia* stem based on mass spectrometric analysis

		MASS						PEAK AREA (N = 3)	
PEAK NO.	RT (MIN)	CALCULATED M/Z	OBSERVED M/Z	Δ ERROR (PPM)	FORMULA	FRAGMENTS M/Z (ABUNDANCE)	IDENTIFICATION	FEMALE	MALE
35.	24.9	294.1125 $[M+H]^+$	294.1114	3.74	$C_{18}H_{15}NO_3$	249.0896 (100), 236.1038 (4), 219.0788 (13), 192.0934 (5)	*N*-formylanonaine[e]	4.9E+03	2.7E+02
36.	45.08	100.0775 $[M+H]^+$	100.0774	1.00	C_5H_9NO	74.0967 (5), 69.0335 (14), 58.0657 (61), 58.0292 (100), 44.0501 (8), 41.0390 (13)	*N*-methyl-2-pyrrolidone[e]	7.0E+02	6.2E+02

[a]guanidino compounds, [b]aporphine alkaloids, [c]benzylisoquinoline alkaloids, [d]protoberberine alkaloids, [e]other *N*-containing compounds, [f]ecdysterone derivatives, [g]sesquiterpenes, F-female, M-male, ‡-authenticated by standards.

Source: Reproduced from Bajpai et al., 2016 with permission from John Wiley and Sons.

The characteristic features of their MS/MS spectra were the presence of dehydrogenated fragments ions, successive dissociation of the substituents, and loss of methyl followed by elimination of H, CH_3 or CO and the absence of product ions below *m/z* 200 (Nair, 2006). Accordingly, the compounds corresponding to the peaks 16, 25, 26 and 28 showed fragmentations involving loss of methyl followed by elimination of H, CH_3, or CO. Peaks 26 and 28 corresponded to jatrorrhizine and palmatine on the basis of authentic standards.

The molecular mass and MS/MS spectrum of the peak 25 was similar to jatrorrhizine; hence, it was tentatively assigned as columbamine, which is isomeric to jatrorrhizine. The MS/MS of peak 16 showed $[M\text{-}CH_3]^+$ ion similar to other protoberberine alkaloids such as jatrorrhizine and columbamine. The shorter retention time of peak 16 indicated that it is more polar than 25 and 26 could be / may be due to the presence of additional OH groups. Exact mass and MS/MS data of peak 16 were corresponding to stepharanine having two additional OH groups each in ring A and D which was reported earlier from *Tinospora capillipes* (Samanani and Facchini, 2001). The absence of $[M\text{-}CH_3\text{-}H]^+$ revealed the lack of vicinal methoxy groups at C9–C10 in the protoberberine skeleton.

Along with protoberberine, some tetrahydroprotoberberine alkaloids were also identified in TCS. The retro-Diels-Alder (RDA) reaction and B-ring cleavage produced characteristic MS/MS fragments for the tetrahydroprotoberberine alkaloids (Nair, 2006). The MS/MS spectra of peaks 9 and 23 showed prominent RDA ions, respectively, at *m/z* 178.0865 and 192.1016, and B-ring cleavage ions, respectively, at *m/z* 151.0740 and 165.0906. Peak 23 corresponded to tetrahydropalmatine as confirmed by comparison with reference standard, and peak 9 was tentatively assigned as tetrahydrojatrorrhizine. Peak 22 produced the most prominent ion at *m/z* 192.1018 by RDA, and a subsequent loss of methyl radical gave the ion at *m/z* 177.0778. The loss of methane along with a methyl radical was not observed in MS/MS spectrum, which indicated the absence of 2, 3-dimethoxy substitution pattern and it was identified as *N*-methyl tetrahydrocolumbamine (Patel, 2011).

2.18 BENZYLISOQUINOLINE ALKALOIDS

In TCS, three compounds (at peaks 7, 10 and 12) showed characteristic fragments of benzylisoquinolines. Peak 7 at *m/z* 272.1271 produced fragment ion at *m/z* 255.0926 due to loss of methylamine and the most prominent benzylic cleavage fragment ion at *m/z* 107.0503 and isoquinoline fragment ion at *m/z* 164.0678. On the basis of the previous reports and recorded MS/MS pattern, peak 7 was identified as the norcoclaurine which is the prototype of the benzylisoquinoline group of alkaloids. Peak 10 was identified as reticuline at *m/z* 330 ($C_{19}H_{23}NO_4$). Loss of CH_3NH_2 (*m/z* 299.1273) followed by CH_3OH (*m/z* 267.1002), the isoquinoline fragment (*m/z* 192.1019) and the benzylic cleavage fragment (*m/z* 137.057) supported the identification (Sarma et al., 1998).

The exact mass and molecular formula (*m/z* 314.1752, $C_{19}H_{24}NO_3^+$) of peak 12 corresponded to that of oblongine. Its MS/MS spectrum showed a prominent ion at *m/z* 58.0663 by RDA fragmentation having a formula $[C_3H_8N]^+$ due to $[(CH_3)_2N{=}CH_2]^+$, thereby showing the presence of two methyl groups on the nitrogen. Elimination of $(CH_3)_2NH$ from $[M]^+$ resulted in the fragment at *m/z* 269.1176, which further loses CH_3OH to yield the fragment at *m/z* 237.0906. Minor fragment ions at *m/z* 175.0763, *m/z* 143.0496, and 145.0620 were supported by available literature (Nair et al., 2006; Peer and Sharma, 1989).

2.19 APORPHINE ALKALOIDS

Aporphine alkaloids showed strong $[M]^{+\cdot}$ or $[M+H]^+$ ions. Aporphine having vicinal hydroxyl and methoxy groups may lose CH_3OH followed by CO. RDA fragmentation of B-ring yielded the ion $[CH_3NH{=}CH_2]^+$ or $[(CH_3)_2N{=}CH_2]^+$ depending on the number of CH_3 groups on the nitrogen. Based on retention time and MS/MS fragmentation, peaks 13 and 15 were identified as magnoflorine and isocorydine respectively, by comparison with reference standards. Based on similar fragmentation, five more aporphine alkaloids as peaks 11, 17, 19, 31 and 33 were, respectively, identified as laurifoline, menisperine, xanthoplainine, isoboldine and boldine. Their MS/MS fragment ions are shown in Table 2.3. Peak 21 was identified as chiloenamine at *m/z* 374.1573 based on exact mass and molecular formula. In MS/MS analysis, fragment ion at *m/z* 58.0655 appeared which clearly indicated the presence of dimethyl substituted tertiary amine. It showed predominant fragment ions at *m/z* 329.1007 due to the loss of dimethylamine. Further loss of methyl radical and HCO_2 appeared at *m/z* 314.0776 and *m/z* 269.0808, respectively. Therefore, the characteristic fragmentations of peak 21 were identified as $[M+H-(CH_3)_2NH]^+$, $[M+H-(CH_3)_2NH-CH_3]^+$, and $[M+H-(CH_3)_2NH-CH_3-CO_2-H]^+$.

2.20 OTHER N-CONTAINING COMPOUNDS

Peak 2 (*m/z* 104.1069) corresponding to $C_5H_{14}NO^+$ showed fragment ion at *m/z* 60.0812 due to protonated trimethyl amine $[C_3H_{10}N]^+$. It was identified as choline after comparing with the standard compound. Peaks 5 and 6 corresponding to elemental composition $[C_5H_{12}NO_2]^+$ and $[C_7H_{14}NO_2]^+$ were identified as quaternary N-containing compounds glycine betaine and proline betaine respectively. The data matched with the earlier MS/MS data for these compounds (Sengupta et al., 2009). Peaks 29 at *m/z* 344.1492 ($C_{19}H_{22}NO_5$) and 30 at *m/z* 314.1378 ($C_{18}H_{20}NO_4$) were, respectively, identified as *N*-trans-feruloyl-4-*O*-methyldopamine and *N*-trans-feruloyltyramine (Nikolic et al., 2012). The MS/MS spectrum of peak 30 showed characteristic fragmentation patterns of ions at *m/z* 177.0538 (acylium ion due to cleavage at the amide bond), 149. 0597 and 145.0284 (further losses of MeOH and CO), 121.0642 ($C_8H_9O^+$), 117.0331 and 89.0391 produced from the ferulic acid portion of the amide. Peak 29 also showed similar fragmentation pattern.

Peaks 35 (*m/z* 294.1114) and 36 (*m/z* 100.0774) were identified as *N*-formylanonaine $[C_{18}H_{15}NO_3]$ and *N*-methyl-2-pyrrolidone $[C_5H_9NO]$, reported from TC showing significant contribution in the immunomodulatory activity. Their MS/MS fragment ions are shown in Table 2.3. *N*-formylanonaine produced fragment ions at *m/z* 249.0896, 219.0788, and 192.0934 due to successive respective losses of CH_3NO, CH_2O and CO.

2.21 ECDYSTERONE AND ITS DERIVATIVES

A number of phytoecdysteroids and their derivatives were previously isolated from TC and reported to contribute to the immunomodulatory activity. The basic chemical structure of ecdysteroids is composed of the cyclopentano-perhydrophenanthrene (sterane) skeleton having a conjugated 7-en-6-one structural system. Ecdysteroids are highly hydroxylated compounds having three to eight hydroxy groups attached to the cyclopentano-perhydrophenanthrene ring. 20-Hydroxyecdysone, ecdysterone and makisterone A were reported

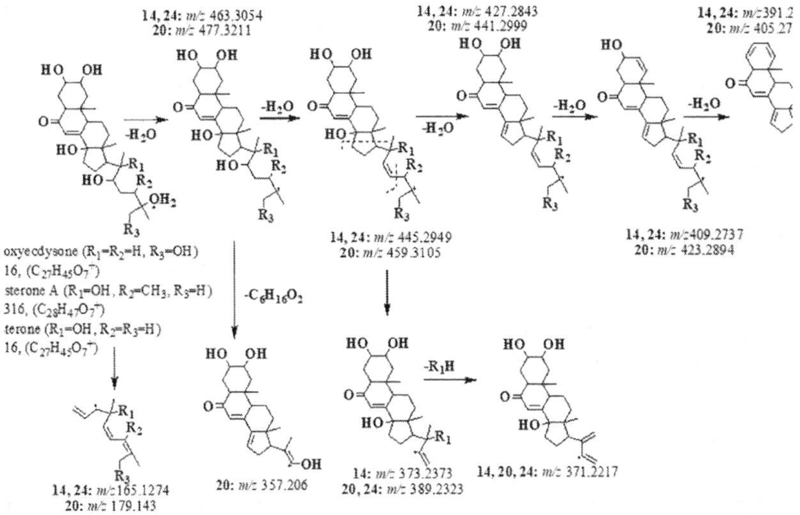

SCHEME 2.1 Proposed fragmentation pathways of peaks 14, 20, and 24 (ecdysterone and derivatives). (Reproduced from Bajpai et al., 2016 with permission from John Wiley and Sons.)

from the TCS. The molecular mass and formulas of peaks 14 and 24 at *m/z* 481.3164 ($C_{27}H_{44}O_7$) and peak 20 at *m/z* 495.3308 ($C_{28}H_{46}O_7$) corresponded to these three compounds, respectively, and their schematic fragmentation pathways are depicted in Scheme 1.

2.22 SESQUITERPENES

Sesquiterpenes, found in plants, are remarkably diverse in terms of their structural properties and have an important role in contributing to the immunomodulatory activity of *T. cordifolia*. Tinocordifolioside, a sesquiterpene glycoside, tinocordifolin and 11-hydroxymustakone, a sesquiterpene ketone are the characteristic compounds of TCS. Compounds detected as peaks 18, 27, and 32 at *m/z* 413.2155, 251.1631, and 235.1692 respectively, generated molecular formula $C_{21}H_{32}O_8$, $C_{15}H_{22}O_3$ and $C_{15}H_{22}O_2$. Peak 18 showed loss of 162 Da from cleavage of glycosidic bond, leading to the ion at *m/z* 251.1623

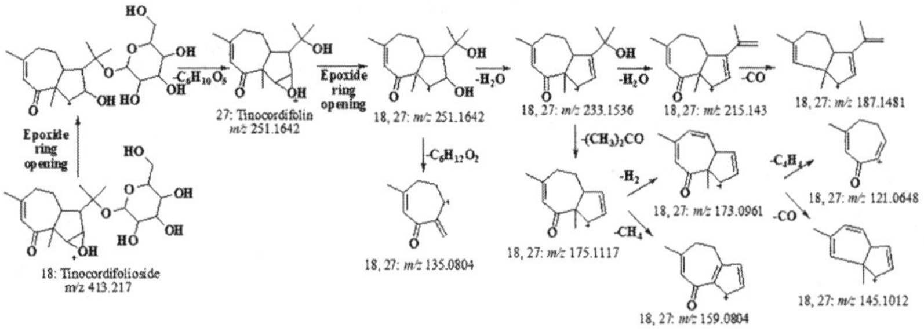

SCHEME 2.2 Proposed fragmentation pathways of peaks 18 and 27 (tinocordifolioside and tinocordifolin). (Reproduced from Bajpai et al., 2016 with permission from John Wiley and Sons.)

corresponding to protonated tinocordifolin (27) which yielded the fragment ion at *m/z* 233.1538. Subsequent loss of acetone produced the most prominent ion at *m/z* 175.1105. The same fragmentations were also observed for peak 27. Hence, peaks 18 and 27 were tentatively identified as tinocordifolioside and tinocordifolin, respectively (Scheme 2). Peak 32 produced fragment ions at *m/z* 217.1629 and 189.1659 due to the loss of H_2O and CO respectively. Ion $[M+H-CO]^+$ yielded fragment ion at *m/z* 147.0803 due to loss of C_3H_6. Based on fragmentation pattern and available literature, peak 32 was tentatively identified as 11-hydroxymustakone.

2.23 PHYTOCHEMICAL VARIATION STUDY IN MALE AND FEMALE SAMPLES

The relative contents of the identified compounds in male and female TCS extracts were compared based on their relative peak area obtained from extracted ion chromatograms. Significant difference was observed in relative concentration due to gender. Significantly (Student's t test $p < 0.05$ or $p < 0.001$) different and higher abundance of magnoflorine, jatrorrhizine, oblongine, γ-guanidino butyl alcohol, and γ-guanidino butyraldehyde were seen in males as compared with females. The mean relative abundances of tetrahydropalmatine, norcoclaurine, and reticuline were significantly ($p < 0.05$ or $p < 0.01$ or $p < 0.001$) higher in females than males (Table 2.4).

TABLE 2.4 Significant difference (t-test) in abundance of major compounds (Mean ± SD, $n = 3$) in male and female *Tinospora cordifolia* stem (TCS)

COMPOUND GROUP	*S. NO.*	*COMPOUND NUMBER*	*IDENTIFIED COMPOUNDS (AS VARIABLES)*	*FEMALE (F) (N = 3)*		*MALE (M) (N = 3)*		*T-VALUE (DF = 4)*	*P VALUE*
				MEAN	*STD. DEV.*	*MEAN*	*STD. DEV.*		
Aporphine alkaloids	1	11	Laurifoline	0.97	0.04	1.21	0.22	−1.85	0.137
	2	13	Magnoflorine	7.06	0.08	9.82	0.25	−18.08	0.000
	3	15	Isocorydine	0.35	0.07	0.43	0.08	−1.24	0.283
	4	17	Menisperine	0.01	0.00	0.01	0.00	0.29	0.789
	5	19	Xanthoplanine	0.01	0.00	0.01	0.00	0.74	0.501
	6	21	Chiloenamine	0.39	0.08	0.54	0.09	−2.06	0.108
	7	31	Isoboldine	0.01	0.00	0.01	0.00	0.43	0.687
	8	33	Boldine	0.02	0.01	0.02	0.00	1.26	0.275
Protoberberine alkaloids	9	9	Tetrahydrojatrorrhizine	3.14	0.38	3.49	0.49	0.96	0.390
	10	16	Stepharanine	0.18	0.01	0.31	0.10	−2.32	0.081
	11	22	*N*-methyl tetrahydrocolumbamine	1.34	0.11	1.61	0.30	−1.48	0.214
	12	23	Tetrahydropalmatine	2.34	0.09	1.82	0.18	4.51	0.011
	13	25	Columbamine	0.93	0.10	0.86	0.09	0.96	0.393
	14	26	Jatrorrhizine	0.73	0.04	1.36	0.34	−3.20	0.033
	15	28	Palmatine	1.98	0.21	2.54	0.37	−2.24	0.088
	16	34	8-Oxojatrorrhizine	0.00	0.00	0.00	0.00	0.87	0.435

(*Continued*)

TABLE 2.4 (*Continued*) Significant difference (*t*-test) in abundance of major compounds (Mean ± SD, $n = 3$) in male and female *Tinospora cordifolia* stem (TCS)

				FEMALE (F) (N = 3)		*MALE (M) (N = 3)*			
COMPOUND GROUP	*S. NO.*	*COMPOUND NUMBER*	*IDENTIFIED COMPOUNDS (AS VARIABLES)*	*MEAN*	*STD. DEV.*	*MEAN*	*STD. DEV.*	*T-VALUE (DF = 4)*	*P VALUE*
Benzylisoquinoline alkaloids	17	7	Norcoclaurine	0.75	0.07	0.46	0.06	5.47	0.005
	18	10	Reticuline	3.17	0.62	1.24	0.45	4.39	0.012
	19	12	Oblongine	0.68	0.09	3.97	0.32	–17.05	0.000
Sesquiterpines/ ecdysterone derivatives	20	14	20-Hydroxyecdysone	1.21	0.34	1.57	0.54	–0.98	0.381
	21	18	Tinocordifolioside	0.21	0.03	0.30	0.06	–2.52	0.065
	22	20	Makisterone A	0.06	0.01	0.03	0.01	2.16	0.096
	23	24	Ecdysterone	0.05	0.01	0.05	0.01	0.46	0.671
	24	27	Tinocordifolin	0.00	0.00	0.01	0.00	–1.92	0.128
	25	32	11-Hydroxy mustakone	0.04	0.01	0.02	0.01	1.46	0.217
Guanidino compounds	26	1	γ-Guanidino butyl alcohol	1.39	0.24	3.81	0.29	–11.20	0.000
	27	3	γ-Guanidino butyric acid	0.23	0.01	0.18	0.05	1.83	0.142
	28	4	γ-Guanidinobutyraldehyde	0.11	0.01	7.30	0.30	–41.31	0.000
	29	8	Arginine	5.63	0.61	1.37	0.09	11.91	0.000

Source: Reproduced from Bajpai et al., 2016 with permission from John Wiley and Sons.

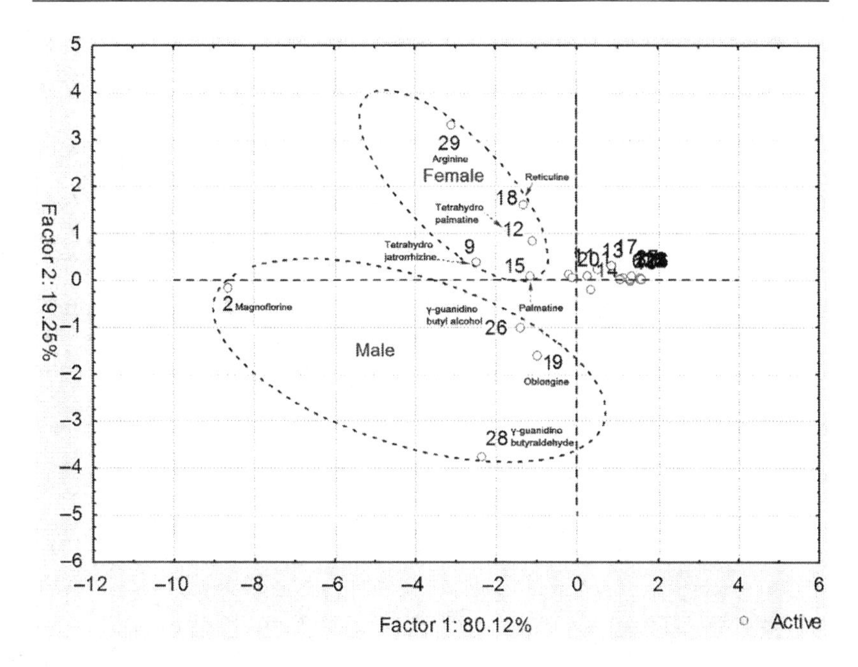

FIGURE 2.4 Projection plot of relative abundance of compounds in male and female *Tinospora cordifolia* stem. (Reproduced from Bajpai et al., 2016 with permission from John Wiley and Sons.)

These findings were also supported by PCA, which differentiated the male and female based on the abundances of compounds (Figure 2.4).

2.24 DISCRIMINATION IN GEOGRAPHICAL SAMPLES

The composition, quality, and quantity of phytochemicals of medicinal plants from different locations and seasons may vary, resulting in variable pharmacological efficacy. Thus, the identification of bioactive phytochemicals of crude herbs and the variations there in according to variations is crucial in order to ensure authenticity, potential efficacy, quality, and safety/conservation of the raw plant material. To identify and validate the possible extent of variations in metabolite profile of TCS collected from different geographical locations and seasons, chemometric analysis was combined with HPLC-ESI-QTOF-MS data. The stock solution of sample and BRS were diluted with methanol to get a range of different concentrations from 50, 100, 250, 500, and 1,000 µg/mL to check the minimum detection response of signal, which was 500 µg/mL of sample and BRS of TCS. All the samples were analyzed in triplicate, and the chemical analysis was based on three independent collections. Total 68 ions were extracted from the base peak chromatogram of geographical and seasonal samples, whereas 54 ions were extracted from gender samples. As direct evaluation of the analytical implication of the results is difficult due to the complexity of spectral data, statistical elaboration of raw data should be implemented to ensure good analytical interpretation and discrimination. Initially, compounds at *m/z* 314.1752, 338.1387, 342.1705, 342.1689 and 352.1546 were selected from HPLC-ESI-QTOF-MS based metabolite profile of TCS because the abundance of these peaks in all geographical samples were most stable and constant. Thus, these peaks have their utility and hence screened for their possible use as markers to predict geographical locations. from the variations in metabolite profile. The mean variances in relative abundance of these compounds were analyzed by ANOVA, and diagnostic accuracy was determined by ROC curve analysis. The mean variance showed that the palmatine has maximum contribution and abundance (Figure 2.5). The data were attested by P-values greater than 0.05. In all the samples, the *t*-test was performed, and their P-value at the 95% confidence interval (CI) demonstrated no systematic error. Sensitivity and Specificity rates for each ion were between 80% and 100% with 95% CI.

The diagnostic accuracy of palmatine was found significant having 100% sensitivity and specificity with 95% CI value as shown in Table 2.5. The analysis from selected compounds showed that palmatine is significant in differentiating the geographical location sample; however, to determine all the possible marker compounds, all the peaks were selected to study variations in geographical samples. Two locations (JBL and LKO) were randomly selected for training set and subjected to FA and PCA to reduce the data dimensionality, which extracted 29 major ions contributing 56.31% of the total variations. The training data set was then subjected to ROC curve analysis to find out the discriminating ability of ions to differentiate samples of two locations JBL and LKO. The ROC curve analysis found 24 significant ($p < 0.05$ or $p < 0.001$) predictor ions for variations in geographical samples, and among these, six ions at *m/z* 130.0494, 264.2301, 279.2323, 282.2796, 294.1139 and 352.1546 (AUC=1.000) were discriminating the samples of two locations (JBL and

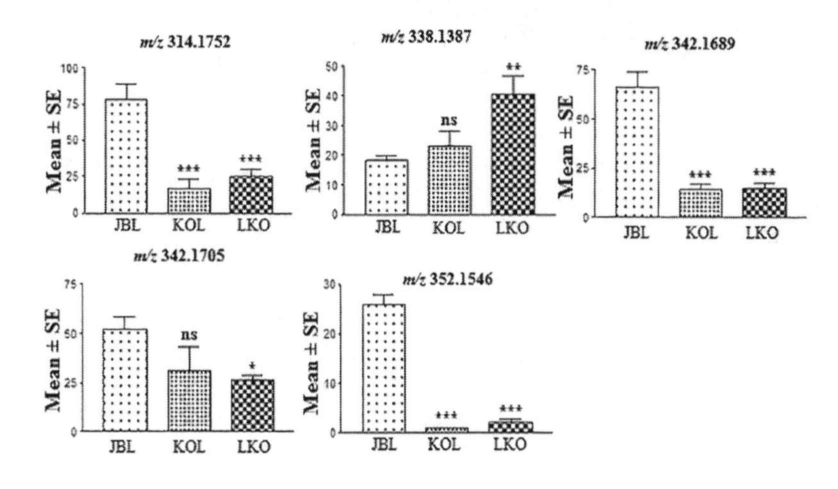

FIGURE 2.5 Relative abundance of five most stable stem constituents of *Tinospora cordifolia* at three different locations: Jabalpur (JBL), Kolkata (KOL), and Lucknow (LKO). (Reproduced from Bajpai et al., 2017b with permission from John Wiley and Sons.)

TABLE 2.5 Diagnostic accuracy of five stable constituents of *Tinospora cordifolia* stem to discriminate Jabalpur and Kolkata samples using receiver operating characteristics curve analysis

M/Z	*CUTOFF VALUE*	*SENSITIVITY (95% CI)*	*SPECIFICITY (95% CI)*	*+PV*	*–PV*
314.1752	≤52.78	100.00 (73.4–100.0)	80.00 (51.9–95.4)	80.0	100.0
338.1387	≤15.4	66.67 (34.9–89.9)	73.33 (44.9–92.0)	66.7	73.3
342.1705	≤10.73	66.67 (34.9–89.9)	100.00 (78.0–100.0)	100.0	78.9
342.1689	≤29.12	100.00 (73.4–100.0)	80.00 (51.9–95.4)	80.0	100.0
352.1546	≤1.15	100.00 (73.4–100.0)	100.00 (78.0–100.0)	100.0	100.0

JBL, Jabalpur; KOL, Kolkata.
Source: Reproduced from Bajpai et al., 2017b with permission from John Wiley and Sons.

LKO) with 100.0% sensitivity (95% CI=78.0–100.0) and 100.0% specificity (95% CI=78.0–100.0).

These 24 predictor ions also successfully discriminated between the test samples of different geographical location; however, among these predictor ions, the two ions at *m/z* 294.1139 and 445.2136 (AUC = 1.000) discriminate the samples of training and test data sets with 100.0% sensitivity (95%

CI = 73.4–100.0) and 100.0% specificity (95% CI = 78.0–100.0). The ROC extracted predictor ions were further subjected to ANOVA to authenticate variations among the geographical locations. For each ion, comparing the relative mean abundance among the groups, ANOVA revealed significantly ($p > 0.001$) different abundance of all ions among the groups. The mean abundance of *m/z* 282.2796 and 318.3017 in JBL was significantly ($p < 0.01$ or $p < 0.001$) higher as compared with both LKO and KOL. The mean abundance of *m/z* 294.1139, 307.1546, and 373.1305 were significantly ($p < 0.001$) higher in both JBL and KOL as compared with LKO. Similarly, the abundance of *m/z* 330.1340 and 445.2362 were significantly ($p < 0.001$) higher in LKO as compared with both JBL and KOL. Moreover, the abundance of *m/z* 359.1501 was significantly ($p < 0.001$) higher in KOL as compared with both JBL and LKO. The variations of abundance and diagnostic accuracy observed in *m/z* 282.2796, 294.1139, 307.1546, 318.3017, 330.1340, 359.1501, 373.1305, and 445.2362 were able to discriminate among the geographical locations and hence may be used as marker peaks to discriminate the geographical location of TCS samples.

2.25 DISCRIMINATION IN SEASONAL SAMPLES

Similar to geographical location, the identification of marker peaks for seasonal variation in TCS, data were randomly divided in the training data set (76% sample) and test data set (24% sample). Training data set was first subjected to FA and PCA, which extracted 23 ions contributing 52.90% of the total variations. The FA/PCA extracted major ions of training data set were further subjected to Mann-Whitney U test to compare the mean relative abundance of two groups (premonsoon and postmonsoon), which revealed significantly ($p < 0.05$ or $p < 0.01$ or $p < 0.001$) different mean abundance of *m/z* 177.0548, 344.1482, 359.1501, 373.1305 and 429.3039.

To confirm the results of FA/PCA, all the 23 extracted ions were subjected to ROC curve analysis. The ROC curve analysis also showed the same five ions, viz., *m/z* 177.0548, 344.1482, 359.1501, 373.1307 and 429.3039 as the significant ($p < 0.01$ or $p < 0.001$) predictor of seasonal variations discriminating the samples of two seasons with 100.0% sensitivity (95% CI = 69.0–100.0) and 100.0% specificity (95% CI = 84.4–100.0). To validate the diagnostic accuracy, these ions were further tested on test data set. However, ions at *m/z* 177.0548 and 429.3039 failed for statistical significance when tested on test data set. To authenticate the prediction ability of the rest three ions at *m/z* 344.1482, 359.1501, and 373.1305, predictor ions were further subjected to *t* test to find out any variations between the seasons. Comparing the relative mean abundance between the two seasons, *t* test revealed significantly ($p < 0.05$ or $p < 0.01$ or $p < 0.001$) different and higher abundance of all ions in premonsoon season as compared with postmonsoon season.

Furthermore, the ROC curve analysis revealed the variation that three significant ions at *m/z* 344.1482, 359.1501 and 373.1305 were contributing maximum to discriminate the seasonal sample ($p < 0.05$ or $p < 0.001$) and among these *m/z* 359.1501 at cut off value of ≤5.18, discriminating the samples of training and test data sets with 100.0% sensitivity (95% CI = 48.0–100.0). The analysis revealed that the variations of abundance and diagnostic accuracy observed in marker ions at *m/z* 344.1482, 359.1501, and 373.1305 were able to discriminate the sample having seasonal variations with 100.0% sensitivity and 100.0% specificity.

2.26 QUANTITATIVE ESTIMATION

The quantitative analysis was accomplished by detection and quantitation controlled by AB Sciex Analyst 1.5.1 software (Applied Biosystems). The UPLC-ESI-MS/MS analysis was performed on a Waters Acquity UPLC™ system (Waters, Milford, MA, USA) interfaced with hybrid linear ion trap triple quadrupole mass spectrometer (API 4000 QTRAP™ MS/MS system from AB Sciex, Concord, ON, Canada) equipped with electrospray (Turbo V™) ion source. The Waters Acquity UPLC™ system was equipped with a binary solvent manager, sample manager, column oven, and photodiode array detector. The compound dependent multiple reaction monitoring (MRM) parameter optimization was performed by direct infusion of 50 ng/mL solutions of the each analyte using a Harvard syringe pump (Harvard Apparatus, South Natick, MA, USA). Quadrupole 1 and quadrupole 2 were maintained at unit resolution.

2.27 UHPLC CONDITIONS FOR QUANTITATION OF ALKALOIDS

Column: Acquity UPLC BEH™ C_{18} column (1.7 μm, 2.1×100 mm)
Mobile phase: (A) 0.1% formic acid aqueous solution (v/v) and (B) acetonitrile
Gradient program: 5% (B) initial to 1.0 min, 5–30% (B) from 1.0 to 2.0 min, hold to 30% (B) from 2.0 to 3.0 min, 30–5% (B) from 3.0 to 4.0 min, and hold back to initial condition from 4.0 to 5 min
Flow rate: 0.3 mL/min
Column thermostate: 30°C
Injection volume: 1 μL
UV detection: 190–400 nm

2.28 ESI-QqQLIT-MS/MS CONDITIONS FOR QUANTITATION OF ALKALOIDS

Scan type: multiple reaction monitoring
Ion source: ESI (positive mode)
Experiment: single period experiment
Optimized source dependent parameters for positive mode
Scan range: *m/z* 100–1,000
Ion Spray Voltage: 5,500 V
Turbo spray temperature: 550°C
Nebulizer gas (GS 1): 50 psi
Heater gas (GS 2): 50 psi
Curtain gas: 20 psi
Collision activated dissociation gas: medium
Interface heater: on
(High purity nitrogen was used for all the processes)

Simultaneous quantitation of selected analytes was carried out using MRM acquisition mode at unit resolution, and its conditions were optimized for each compound during infusion. The developed UPLC-(+ESI)-MS/MS method for quantitative analysis in the MRM mode was validated according to the guidelines of the International Conference on Harmonization (2005) for the simultaneous quantitative determination of quaternary protoberberine and aporphine alkaloids by determining calibration curves, limit of detection (LOD), limit of quantitation (LOQ), specificity, precision, solution stability and recovery.

The linearity of the calibration was performed by the analytes-to-IS peak area ratios versus the nominal concentration. The calibration curves were constructed using eleven different concentrations with a weight ($1/x^2$) factor by least squares linear regression analysis. For each reference compound, the linearity of calibration curve was determined based on five experiments. The LODs and LOQs were measured by signal-to-noise ratio (S/N) of 3 and 10, respectively. The linearity was evaluated by the coefficient of determination (r^2) of the respective calibration curves, which showed r^2 within the test ranges (0.99966 to 0.99988). LOD and LOQ for each analyte varied from 0.07–0.24 ng/mL and 0.21–0.74 ng/mL, respectively.

Relative standard deviation (RSD) values for precision were in the range of 0.78–2.04% for intraday assays and 0.75–2.06% for interday assays. RSD for stability was analyzed by replicating injections of the sample solution at 0, 2, 4, 8, 12, and 24 hours and found in the range of 0.79–2.12%. Three spike levels were set at 50%, 100%, and 200% of each reference standard for the recovery test. RSD for recovery was found in the range of 1.05–2.64%. The recoveries of the compounds evaluated were 96.93–101.52% ($n = 3$) by calculating the ratio of amount detected versus the amount added.

The five bioactive alkaloids of TCS were analyzed in three different pooled extracts of male and female of 2012, 2013, and 2014 (Table 2.6). It was observed that magnoflorine was the most abundant compound in TCS, and it was 8.10 mg/g in male and 3.85 mg/g in female stem. Other alkaloids such as palmatine, jatrorrhizine, and isocorydine were also more in male stem, whereas tetrahydropalmatine was an exception, which was marginally higher in female stem. The effect of gender (male and female) on abundance of analytes using ANOVA revealed significant effect of both analytes ($F = 119.86$, $p < 0.001$) and plant sex ($F = 46.35$, $p < 0.001$). The interaction effect of analytes and plant sex was also found significant ($F = 19.81$, $p < 0.001$). The mean content of analytes magnoflorine, jatrorrhizine and palmatine was higher in males. However, the mean abundance of other analytes did not differ ($p < 0.05$) between the two groups and found to be statistically the same. Overall (average of male and female), the mean abundance of magnoflorine was found significantly ($p < 0.001$) different and higher as compared with other analytes.

TABLE 2.6 Content of selected analytes (mg/g) in male and female in *Tinospora cordifolia* stem

SAMPLE	*JATRORRHIZINE*	*MAGNOFLORINE*	*ISOCORYDINE*	*PALMATINE*	*TETRAHYDROPALMATINE*	*TOTAL (MG/G)*
TCS-Male 12	1.71	8.10	1.01	2.11	0.61	13.54
TCS-Male 13	1.52	6.30	0.94	2.26	0.49	11.51
TCS-Male 14	1.63	6.75	0.83	2.05	0.32	11.58
TCS-Female 12	1.62	3.85	0.84	1.14	0.93	8.38
TCS-Female 13	1.31	2.94	0.20	1.89	0.59	6.93
TCS-Female 14	1.17	3.16	0.16	0.75	0.26	5.50

T. cordifolia stem (TCS) male and female collected in year 2012, 2013, and 2014 (12, 2012; 13, 2013; 14, 2014).
Source: Reproduced from Bajpai et al., 2016 with permission from John Wiley and Sons.

3 Correlative Pharmacological Activity of *Tinospora cordifolia*

Bioactivity of *Tinospora cordifolia* (TC) is correlated with the presence of alkaloids (Saha et al., 2012). Alkaloids are the active phytoconstituents responsible for the anticancer activity in this plant (Panchabhai et al., 2008; Singh et al., 2017). A phytoecdysteroid, 20-hydroxyecdysone, has been also reported to be useful in plant physiology, plants defense against insects, and pharmaceutical applications (Goel et al., 2004; Chandrasekaran et al., 2009). Furthermore, choline, jatrorrhizine, magnoflorine, isocorydine, palmatine, tetrahydropalmatine and 20-hydroxyecdysone are reported to exhibit multiple biological activities such as antiadipogenic, antidiabetic, cytotoxicity, antifungal, cardioprotective, anticancer and anti human immuno deficiency virus (HIV) (Jagetia and Rao, 2006b; Reddy et al., 2009; Singh et al., 2017). The TC leaves showed antidiabetic action in alloxanized diabetic rabbit. TC root extract exhibited lowering of alkaline phosphatase, hepatic glucose-6-phosphatase, lactate level and serum acid phosphatase level in diabetic rats causing the hypoglycemic and hypolipidemic effects (Stanely et al., 2000). The antibacterial and antiulcer activities in TC roots are due to butyryl cholinesterinase inhibitory activity caused by the presence of protoberberine alkaloids jatrorrhizine in roots (Sarma et al., 1995; Sarma et al., 1996; Moody et al., 2006; Sengupta et al., 2009). TC extract protects carbon tetrachloride intoxication in mice (Bishayi et al., 2002). TC is used as hypoglycemic agent, and its commercial herbal preparations showed competitive inhibition activities of α-glucosidase and sucrase (Sengupta et al., 2009). TC has also shown the antiosteoporotic effects, as it contains ecdysteroids such as 20-hydroxyecdysone and β-ecdysone that has antiosteoporotic property and the antiosteoporotic effect of β-ecdysterone does not involve the activation of the estrogen receptor (Kapur et al., 2010; Seidlova-Wuttke et al., 2010). The immunostimulant activity of TC was due to the presence of diterpenoid glycosides such as cordifolioside A and cordifolioside B (Murya et al., 1996). The macrophage activation in immune responses by TC extract was due to occurrence of another diterpenoid glycoside cordial (Kapil and Sharma, 1997; Nair et al., 2006). Chitin and tinosporin from TC plant targeted viruses (retroviruses) (HIV-1, HIV-2) of all subgroups, human T-lymphotropic virus and other viral disease (Chen et al., 2009; Chetan and Nakum, 2010). The uterine contraction activity of TC plant was observed due to the protoberberine alkaloid palmatine (Iwasa et al., 1998).

Aporphine alkaloid such as magnoflorine is reported to suppress the induction of cellular immune response (Devasagayam and Sainis, 2002) and also as a hypotensive agent, which inhibits lipid oxidation in high density lipoprotein. Isoquinoline alkaloids such as magnoflorine, jatrorrhizine and palmatine reported from the alkaloidal fraction of TC stem (TCS) have shown in vitro and in vivo hyperglycemic effect (Peer and Sharma, 1989; Wadood et al., 1992). The free radical scavenging activity of TCS is due to the alkaloids such as magnoflorine, palmatine, tetrahydropalmatine and tinosporin (Panchabhai et al., 2008). Epicatechin, a flavonoid present in the TCS, reported to have in vitro antioxidant activity (Pushp et al., 2013). Secoisolariciresinol, a glycoside, obtained from TC also showed antioxidant activity in 2,2-diphenyl-1-picryl-hydrazyl-hydrate assay (Singh et al., 2017). The ethanolic extracts of TCS are described to cause a reduction in blood glucose level, which specifies certain indirect action of the TC herbal drug on carbohydrate metabolism, implying that the herbal drug favorably affects the insulin secretion and shows the inhibition of peripheral glucose release. A polysaccharide isolated from TCS showed inhibition of metastases in the mice (Khaleque et al., 1971; Kapil and Sharma, 1997; Jahfar, 2003; Jagetia and Baliga, 2004; Jagetia and Rao, 2006b; Kapur et al., 2010). The chemopreventive ability against human colon cancer was also reported from a furanolactone diterpenoid compound columbine isolated from the TCS (Kohno et al., 2002). TC extract led to the production of antitumor molecules such as IL-1 and reactive nitrogen intermediates, and tumor necrosis factor was observed along with significant cytotoxicity. The immunomodulatory activity in TC was also reported from tinocordiside, a sesquiterpene glycoside isolated from stem (Ghosal et al., 1997; Geeta et al., 2007; Bala et al., 2015b). A dose dependent antiinflammatory activity of TCS was also detected due to columbin (Singh et al., 2003b; Sharma et al., 2010a; Sharma et al., 2010b). The aqueous fraction of TCS is rich in polysaccharide and shown to be active in reducing the metastatic potential of melanoma cells (B16F-10) (Leyon and Kuttan, 2004). The TCS has a preventive effect against chemically induced hepatocellular carcinoma in rats due to the presence of a clerodane diterpenoid (5R, 10R)-4R, 8R-dihydroxy-2S, 3R:15,16-diepoxycleroda-13(16), 17,12S:18,1S-dilactone (Dhanasekaran et al., 2009).

3.1 EVALUATION OF IMMUNE RESPONSE

The activity evaluation of TCS was carried out on BALB/c mice (18–20 g) inbred strain with matched age and sex. The animals were housed under standard conditions of temperature (23 ± 1°C), relative humidity (55 ± 10%), and light/dark cycles (12 h/12 h) at National Laboratory Animal Centre of CSIR-Central Drug Research Institute, Lucknow, India, and fed with standard pellet diet and water *ad libitum*. The crude extracts of male and female TCS and standard levamisole were prepared as suspension in distilled water and fed once daily to the animals for 14 consecutive days using canula. The control animals received only water in the same manner. For test samples, each treatment group comprised five animals in triplicate, and for standard and control, one group of five animals were used. The daily dose of plant samples and standard was, respectively, 3, 10, 30, and 3 mg/kg.

3.2 REACTIVE OXYGEN SPECIES

The peritoneal exudates cells (PECs) were collected from treated and control mice to determine the oxidative burst in the antigen presenting cells by fluorometric assay using dichlorofluorescin diacetate (DCF-DA). Briefly, freshly harvested PECs of both treated and untreated groups were adjusted to a concentration of 1×10^6 cells/mL in phosphate buffer saline (PBS), washed thrice with PBS, and transferred to fluorescence activated cell sorting (FACS) tubes (1×10^6 cells/tube). For probe loading, cells were incubated with the 2′, 7′-dichlorodihydrofluorescein diacetate (H2DCF-DA) at a final concentration of 1 μM, for 15 min at 37°C in dark and washed twice in PBS.

The nonfluorescent H2DCF-DA is converted to the highly fluorescent 2′, 7′- DCF by oxidation and reactive oxygen species (ROS) levels in individual living cells were determined by measuring their fluorescence intensity on FACS Calibur (BD, San Diego, CA, USA). Data were analyzed by CellQuest Software (BD, San Diego, CA, USA), and mean ROS values were evaluated for cell populations.

3.3 CD4/CD8 T AND CD19 B LYMPHOCYTE POPULATION

Splenocyte suspension was prepared from spleens of treated and untreated control mice on day 15 of the commencement of treatment. 1×10^6 cells were first blocked with Mouse Seroblock FcR at room temperature (RT) for 10 min, washed and labelled with rat antimouse fluorescein isothiocyanate cluster of differentiation 4 (FITC-CD4) monoclonal antibody for 10 min at RT and reincubated with rat antimouse peritoneal exudate cluster of differentiation 8(PE-CD8) for another 10 min. The other set of tubes were used to analyze CD19 cells after incubation for 10 min, with rat antimouse FITC-CD19. Cell pellets were suspended in sheath fluid and analyzed by FACS using CellQuest analysis software after gating the forward and side scatter settings to exclude debris. For each determination 10,000 cells were analyzed, and the results are reported as percentage of each cell population.

3.4 INTRACELLULAR CYTOKINES

Splenocytes (4×10^6/mL) were incubated with brefeldin A (10 μg/mL, BD) in dark for 6 h in CO_2 incubator at 37°C and reincubated with mouse Seroblock FcR at RT for another 10 min. Cells were washed in PBS and incubated with FITC rat antimouse CD4 antibody. Leucoperm A and Leucoperm B were added at RT for 15 min, and cells were dispensed in four tubes each containing 1×10^6 cells/100 μL. PE rat antimouse monoclonal antibodies to cytokines interleukin-2 (IL-2), interferon gamma (IFN-γ), interleukin-4 (IL-4) and interleukin-10 (IL-10) were added to separate tubes. Cells were washed and finally suspended in 250 μL of PBS for FACS readings. For each determination

10,000 cells were analyzed, and the results are reported as percentage of each cell population after gating based on isotype controls.

Biological activity data were expressed as the mean ± standard error (S.E.) and statistical analysis was carried out by one-way analysis of variance (Dunnett's multiple comparison test). A conventional $p < 0.05$ was taken as evidence of significant differences, $p < 0.01$ and $p < 0.001$ were considered as highly significant, whereas $p > 0.05$ was not significant. Within an experiment, each data point was the average of triplicates in a single experiment.

3.5 IN VIVO EVALUATION OF IMMUNE RESPONSE

Oral administration of crude stem extracts of TC male and female plants at 3, 10, and 30 mg/kg doses to BALB/c mice for 14 consecutive days led to significant expansion of CD4+ T helper cell and CD8+ T cytotoxic (Tc) cell population in mice spleen in a dose dependent manner ($p < 0.05$–$p < 0.001$). Highest cell proliferation was observed at the highest dose of 30 mg/kg (Figure 3.1a and b). Changes in CD19 B cell population followed a dose dependent rise, and the increase was statistically significant only in the case of female plant at 10 and 30 mg/kg ($p < 0.05$–$p < 0.001$) and could be correlated with 3 mg/kg dose of levamisole ($p < 0.01$) (Figure 3.1c). In vitro oxidative burst in the peritoneal cells of mice was significantly ($p < 0.001$) induced by female crude extract in a dose dependent manner at all the doses as compared with untreated controls (Figure 3.2a). A similar trend was also observed with the male plant extract; however, the immune response was statistically not significant ($p > 0.05$) except at 30 mg/kg ($p < 0.001$) (Figure 3.2a). Levamisole also induced significant ($p < 0.001$) ROS production from peritoneal cells as compared with control group (Figure 3.2a).

The intracellular production of both type 1 T helper (Th1) and type 2 T helper (Th2) cytokines was significantly upregulated in the CD4 T cells in the spleen isolated from mice fed with the male and female stem extracts ($p < 0.05$–$p < 0.001$). Dose dependent increase in the production of IL-2 was significant in the female stem extract treated mice at 10 mg/kg ($p < 0.05$) and

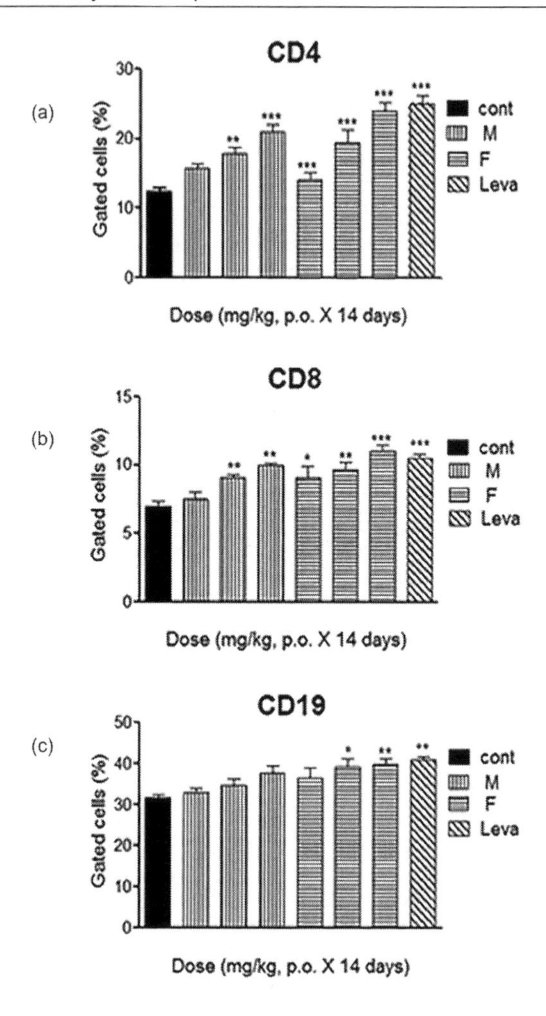

FIGURE 3.1 Flow cytometric analysis of T and B cell surface antigens in BALB/c mice: (a) CD4 T cells; (b) CD8 T cells; (c) CD19. Bars represent mean ± SE; *$p < 0.05$ low significant; **$p < 0.01$ as highly significant; ***$p < 0.001$ very highly significant (Cont, control group; M, male plant extract group; F, female plant extract group; Leva, levamisole group). (Reproduced from Bajpai et al., 2017a with permission from Elsevier.)

30 mg/kg ($p < 0.001$) akin to standard levamisole ($p < 0.001$); however, the male stem extract could not yield significant response (Figure 3.2b). Another Thl cytokine, IFN-γ, also revealed significant dose dependent increases at 10 mg/kg ($p < 0.001$) and 30 mg/kg ($p < 0.001$) in mice fed with female stem

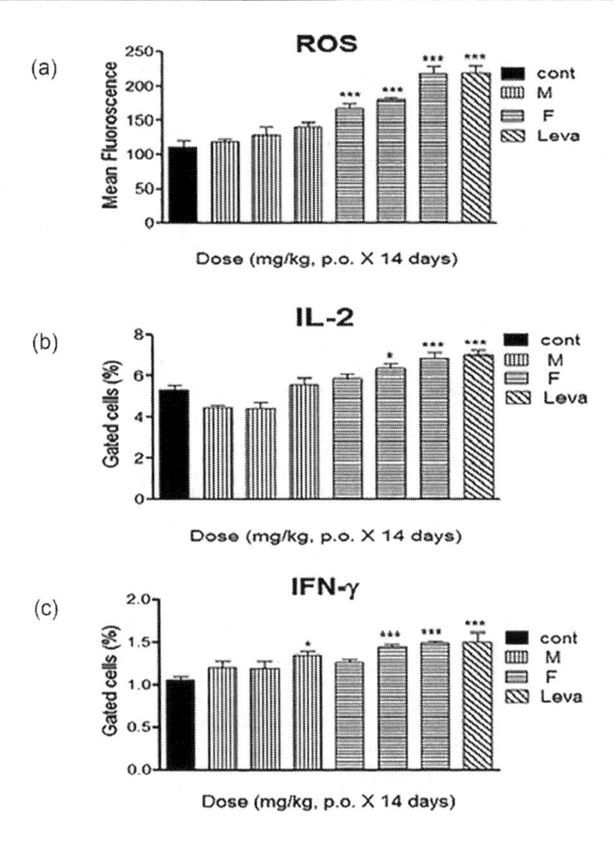

FIGURE 3.2 Flow cytometric analysis of oxidative burst in peritoneal macrophages and IL-2 and IFN-γ intracellular cytokine in CD4 T cells using splenocytes of BALB/c mice: (a) ROS; (b) IFN-γ; (c) IL-2. Bars represent mean ± SE; $*p < 0.05$ low significant; $**p < 0.01$ highly significant; $***p < 0.001$ very highly significant (Cont, control group; M, male plant extract group; F, female plant extract group; Leva, levamisole group). (Reproduced from Bajpai et al., 2017a with permission from Elsevier.)

extract, whereas a significant rise was only at the highest dose of 30 mg/kg ($p < 0.05$) in case of male stem extract (Figure 3.3c). Levamisole exerted similar effects and increased the percentage of IFN-γ producing CD4+ T cells ($p < 0.001$) (Figure 3.3c). As in case of levamisole, the female stem extract induced the production of Th2 cytokine IL-4 at a higher dose of 30 mg/kg ($p < 0.001$); however, the male stem extract showed a significant increase at a lower level of 3 mg/kg ($p < 0.05$). It was interesting to note that both male and female test samples augmented the production of antiinflammatory cytokine

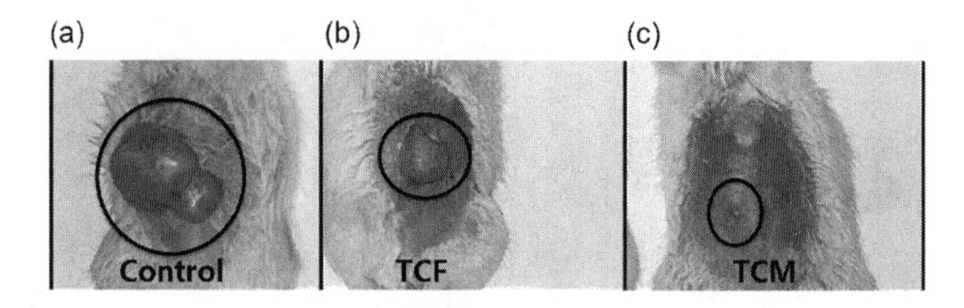

FIGURE 3.3 *In vivo* antitumor efficacy of *Tinospora cordifolia* male (TCM) and *Tinospora cordifolia* female (TCF) in syngeneic rat LA7 mammary tumor model (a) Control; (b) Treatement of tumor with Tinospora cordifolia female (TCF) stem extract; (c) Treatement of tumor with Tinospora cordifolia male (TCM) stem extract.

IL-10 at all the three doses ($p < 0.001$); however, between the two extracts, female extract generated better response especially at 30 mg/kg, which was at par with levamisole.

3.6 EVALUATION OF CELL CYTOTOXICITY

The cells (5×10^3 cells/well) were seeded in 96 well plates in medium containing 10% fetal bovine serum (FBS) and grown till 60% of confluency. The cells were treated with TCS extract dissolved in dimethyl sulfoxide (DMSO) with final concentration of 0.01% and diluted to appropriate concentrations with culture media containing 2% charcoal stripped FBS for another 24 h. Vehicle (DMSO) at final concentration of 0.01% in culture media containing 2% FBS was treated as negative control. 3-(4,5-Dimethylthiazol-2-yl)-2, 5-diphenyltetrazolium bromide (MTT) was added to each well to a final concentration of 0.5 µg/µL in each well and incubated for 4 h. The formazan crystals were dissolved in 100 µL/well of DMSO by keeping the 96 well plates on plate shaker (PSU-2T Minishaker, Biosan) at RT for 5 min, and absorbance was recorded at 540 nM with a plate reader (MQX-200, Bio-TEK Instruments Inc). The cell percent inhibition was calculated from the absorbance of treated versus negative control and Prism 3.0 version (Graphpad Software, USA) was used to calculate the IC_{50} values of the treated groups. The treatment group was assayed in triplicate wells in each experiment, and triplicate experiments were conducted.

TCS has been shown to exhibit anticancer activities. In an earlier study in human breast cancer cells the ethanolic extracts of TCS has shown a dose dependent cytotoxic effect, whereas the aqueous extracts failed to induce significant anticancerous activity (Maliyakkal et al. 2013). TC like any other plants due to its sessile and perennial nature is subjected to a number of factors from soil to climate and altitude. How these factors impact the drug quality of the plant has remained largely unexplored. Work done on *Piper betle* has shown the difference in chemical constituents due to gender may affect the therapeutic potential of the plant material (Bajpai et al., 2012).

However, till date, there are no reports to validate whether the TCS exhibits any differential activity due to edaphic factors (location of collection site) and the phenomenon of dioecy in this plant.

To fill this gap, the evaluation of TCS extract from different geographical locations and gender against breast cancer cell lines such as Michigan Cancer Foundation-7 breast cancer model (MCF-7; nonmetastatic breast cancer model), M.D. Anderson metastasis breast cancer 231 model (MDA-MB-231; aggressive breast cancer model) and prostate cancer cell lines DU-145 and prostate cancer 3 (PC-3) cell line along with human embryonic kidney (HEK-293) cells using MTT assay to assess cell cytotoxicity, were undertaken. The results are shown in Table 3.1.

As evident from the results, TCS male (TCSM) extract was most efficient in inhibiting the growth of cancerous cell lines with the reported IC_{50} of 30.27 ± 2.44 µg/mL in MCF-7 cell line (Figure 3.3), 57.1 ± 3.67 µg/mL in MDA-MB-231 cell line, 39.1 ± 0.81 µg/mL in DU-145 cell line and 21.18 ± 2.02 µg/mL in PC-3 cell line with the value of IC_{50} in HEK-293 cell line >100. The effect of TCSM on the percent cell cytotoxicity in different cell lines such as MCF-7, MDA-MB-231, PC-3, DU-145 and HEK-231 is presented in Table 3.1.

Among the extracts screened, TCS female (TCSF) extract showed >100 IC_{50} in all the breast cancer cell lines along with HEK-293. TCSF extract was also found to be active in inhibiting the cell proliferation in the prostate cancer cell lines; the IC_{50} was 52.44 ± 1.79 µg/mL in DU-145 cell line and 3.71 ± 1.35 µg/mL in PC-3 cell line. TCS Lucknow (TCSL) was active in DU-145 with the IC_{50} of 36.7 ± 1.69 and 56.06 ± 3.01 in PC-3, TCS Kolkata (TCSK) was active in DU-145 with the IC_{50} 37.54 ± 1.12 and TCS Jabalpur (TCSJ) was active against PC-3 with the IC_{50} of 8.954 ± 1.27 (Table 3.1).

TABLE 3.1 Evaluation of *Tinospora cordifolia* stem extract on different cancer cell lines

		ANTICANCER ACTIVITY IC_{50} (μg/ML) IN CELL LINES				
ENTRY	*SAMPLE*	*MCF-7*	*MDA-MB-231*	*HEK-293*	*DU-145*	*PC-3*
1	TCS MALE	30.27 ± 2.44	57.1 ± 3.67	>100	39.1 ± 0.81	21.18 ± 2.02
2	TCS FEMALE	104 ± 2.47	324.09 ± 2.16	>100	52.44 ± 1.79	3.71 ± 1.35
3	TCSL	86.66 ± 1.5	102.5 ± 5.60	30.03 ± 2.8	36.7 ± 1.69	56.06 ± 3.01
4	TCSK	82.69 ± 2.49	99.8 ± 4.70	32.59 ± 1.95	37.54 ± 1.12	77.42 ± 2.52
5.	TCSJ	91.22 ± 2.06	403 ± 8.62	>100	89.95 ± 4.9	8.954 ± 1.27

Source: Reproduced from Bajpai et al., 2017b with permission from John Wiley and Sons.

3.7 MORPHOLOGICAL ASSAY AND EVALUATION

The MDA-MB-231 cells were seeded in the 6 well culture plates with the cell density of 1×10^5 per well and grown up to 80% of cell confluency in low-glucose Dulbecco's Modified Eagle's Medium supplemented with 10% FBS. To identify the impact of TCS (M) on invasive breast cancer, the dose dependent effect of TCS (M) on MDA-MB-231 cells was analyzed by treating the cells at the sub-IC_{50} doses of TCS (M) at 0.5, 1, 10 and 25 µg/mL for 24 h. The phase contrast microscopy was performed at 12 and 24 h. The results showed no significant change in cell morphology at 10 µg/mL up to 12 h; however, the number of floating cells were increased with 25 µg/mL. The results at 24 h also revealed that the cell morphology was significantly affected in a dose dependent manner as compared with the control.

4 Conclusion

Tinospora cordifolia (TC) stem is one of the most widely used plants in Ayurvedic medicine. It balances the vital body fluids and protects from diseases. In terms of consumption, it ranks among the top five plants in Indian system of medicine and is used for its many biological properties. The variation in the bioactivity of this plant is directly related with the presence of diverse phytochemicals. Thirty six phytochemicals including alkaloids, sesquiterpenes and phytoecdysteroids in TC stem, were identified and characterized using high-performance liquid chromatography–electrospray ionization–quadrupole time of flight mass spectrometry and tandem mass spectrometry techniques. Ultra-performance liquid chromatography–electrospray ionization–quadrupole linear ion trap mass spectrometry methods were found to be effective for quantitation of phytochemicals in TC stem. Phytochemicals in male and female TC stem showed significant quantitative variations. It is also evident that female TC stem showed better immunomodulatory efficacy in terms of dose dependent response and superior immune stimulation. It was more efficient in inducing the T cell response via production of both pro inflammatory and anti-inflammatory cytokines. The anticancer bioactivity evaluation revealed cytotoxicity exhibited by the male and female plants; however, male plant showed better anticancer activity. These findings underscore the importance of gender in all the dioecious medicinal plants, where vegetative parts are used. Thus the pharmacological activity may vary depending on the sex of the plant. In the light of these findings, the importance of gender becomes very important in dioecious medicinal plants where only vegetative parts (fruits and seeds are of no consequence due to their being a negligible fraction) are being used as drug. Therefore, while dealing with dioecious medicinal plants, it will be useful to evaluate the bioactivity based chemical profile of the two sexes. This will ensure better efficacy and also reduce the variations, which are very common with plant based drugs. These methods have potential use in the quality control of TC and screening of herbal preparations.

References

Acharya Y. T. *Sushruta Samhita of Sushruta; Sutrasthana; Bhumipravibhagiya*. Reprinted; Chaukhamba Sanskrita Sansthana, Varanasi, India, 2010.

Acharya Y. T. *Charaka Samhita of Agnivesha; Kalpasthana; Madanakalpa*. Reprinted; Chaukhamba Sanskrita Sansthana, Varanasi, India, 2011.

Ahmad F., Ali M. and Alam P. "New phytoconstituents from the stem bark of *Tinospora cordifolia* Miers." *Natural Product Research* 24, no. 10 (2010): 926–934.

Ahmed F. A., Bristy R. S. and Tasnova N. J. "Ethnomedicinal practice of *Tinospora cordifolia* (Willd.) Meirs ex Hook f. & Thoms. by the traditional medicine practitioners at Savar, Dhaka." *Jahangirnagar University Journal of Biological Sciences* 4, no. 2 (2015): 47–51.

Akula R. and Ravishankar G. A. "Influence of abiotic stress signals on secondary metabolites in plants." *Plant Signaling & Behavior* 6, no. 11 (2011): 1720–1731.

Anonymous. The Ayurvedic Pharmacopoeia of India. Published by department of AYUSH, ministry of health and family welfare; Government of India, New Delhi, 1, no. 1 (1999): 53–55.

Asthana J. G., Jain S., Ashutosh M. and Vijay Kanth M. S. "Evaluation of antileprosy herbal drug combinations and their combinations with dapsone." *Indian Drugs* 38, no. 2 (2001): 82–86.

Bajpai V., Pandey R., Negi M. P. S., Kumar N. and Kumar B. "DART MS based chemical profiling for therapeutic potential of *Piper betle* landraces." *Natural Product Communications* 7, no. 12 (2012): 1627–1629.

Bajpai V., Singh A., Chandra P., Negi M. P. S., Kumar N. and Kumar B. "Analysis of phytochemical variations in dioecious *Tinospora cordifolia* stems using HPLC/QTOF MS/MS and UPLC/QqQLIT-MS/MS." *Phytochemical Analysis* 27, no. 2 (2016): 92–99.

Bajpai V., Kumar S., Singh A., Bano N., Pathak M., Kumar N., Misra-Bhattacharya S. and Kumar B. "Metabolic fingerprinting of dioecious *Tinospora cordifolia* (Thunb) Miers stem using DART TOF MS and differential pharmacological efficacy of its male and female plants." *Industrial Crops and Products* 101 (2017a): 46–53.

Bajpai V., Kumar S., Singh A., Singh J., Negi M. P. S., Bag S. K., Kumar N., Konwar R. and Kumar B. "Chemometric based identification and validation of specific chemical markers for geographical, seasonal and gender variations in *Tinospora cordifolia* stem using HPLC-ESI-QTOF-MS analysis." *Phytochemical Analysis* 28, no. 4 (2017b): 277–288.

Bala M., Pratap K., Verma P. K., Singh B. and Padwad Y. "Validation of ethnomedicinal potential of *Tinospora cordifolia* for anticancer and immunomodulatory activities and quantification of bioactive molecules by HPTLC." *Journal of Ethnopharmacology* 175 (2015a): 131–137.

Bala M., Verma P. K., Awasthi S., Kumar N., Lal B. and Singh B. "Chemical prospection of important ayurvedic plant *Tinospora cordifolia* by UPLC-DAD-ESI-QTOF-MS/MS and NMR." *Natural Product Communications* 10, no. 1 (2015b): 43–48.

Bhide B. V., Phalinkar N. L. and Pranjape K. "Chemical Investigation of *Tinospora cordifolia* (Miers)." Journal of The University of Bombay 10 (1941): 89–92.

Bishayi B., Roychowdhury S., Ghosh S. and Sengupta M. "Hepatoprotective and immunomodulatory properties of *Tinospora cordifolia* in CCl4 intoxicated mature albino rats." *The Journal of Toxicological Sciences* 27, no. 3 (2002): 139–146.

Bramwell D. "How many plant species are there." *Plant Talk* 28 (2002): 32–34.

Chadha Y. R. "The Wealth of India, Publication and Information Directorate." *CSIR: New Delhi* 33 (1948): 251.

Chandrasekaran C. V., Mathuram L. N., Daivasigamani P. and Bhatnagar U. "*Tinospora cordifolia*, a safety evaluation." *Toxicology in Vitro* 23, no. 7 (2009): 1220–1226.

Chen J-H., Du. Z-Z, Shen Y-M. and Yang Y-P. "Aporphine alkaloids from *Clematis parviloba* and their antifungal activity." *Archives of Pharmacal Research* 32, no. 1 (2009): 3–5.

Chen S-L., Yu H., Luo H-M., Wu Q., Li C-F. and Steinmetz A. "Conservation and sustainable use of medicinal plants: problems, progress, and prospects." *Chinese Medicine* 11, no. 37 (2016): 1–10.

Chetan B. and Nakum A. "Use of natural compounds, chitin and tinosporin for the treatment of the targeted viruses (retroviruses) (HIV-1, HIV-2) all subgroups, HTLV and other viral disease." Indian Patent Appl (2010): IN 2010MU01350 A 20100730.

Choudhary N., Siddiqui M. B. and Khatoon S. "Pharmacognostic evaluation of *Tinospora cordifolia* (Willd.) Miers and identification of biomarkers." *Indian Journal of Traditional Knowledge* 13, no. 3 (2014): 543–550.

Chulet R. and Pradhan P. "A review on rasayana." *Pharmacognosy Reviews* 3, no. 6 (2009): 229.

de Villiers A., Venter P. and Pasch H.. "Recent advances and trends in the liquid-chromatography–mass spectrometry analysis of flavonoids." *Journal of Chromatography A* 1430 (2016): 16–78.

Devasagayam T. P. A. and Sainis K. B. "Immune system and antioxidants, especially those derived from Indian medicinal plants." *Indian Journal of Experimental Biology* 40, no. 6 (2002): 639–655.

Dhanasekaran M., Baskar A-A., Ignacimuthu S., Agastian P. and Duraipandiyan V. "Chemopreventive potential of Epoxy clerodane diterpene from *Tinospora cordifolia* against diethylnitrosamine-induced hepatocellular carcinoma." *Investigational New Drugs* 27, no. 4 (2009): 347–355.

Dipali G., Vijender S. and Dhan P. "Chemical constituents from *Tinospora cordifolia* stem." *Journal of Medicinal and Aromatic Plant Sciences* 31, no. 3 (2009): 209–214.

Dwivedi S. K. and Enespa A. "*Tinospora cordifolia* with reference to biological and microbial properties." *International Journal of Current Microbiology and Applied Sciences* 5, no. 6 (2016): 446–465.

Gargi M., Thakur S., Anand S. S., Choudhary S. and Bhardwaj P. "Development and characterization of genomic microsatellite markers in *Tinospora cordifolia*." *Journal of Genetics* 96, no. 1 (2017): 25–30.

Geetha K. A., Josphin M. and Maiti S. "Gender instability in *Tinospora cordifolia*: an immunomodulator." *Current Science* 92, no. 5 (2007): 591–592.

Ghosal S. and Vishwakarma R. A. "Tinocordiside, a new rearranged cadinane sesquiterpene glycoside from *Tinospora cordifolia*." *Journal of Natural Products* 60, no. 8 (1997): 839–841.

Goel H. C., Prasad J., Singh S., Sagar R. K., Agrawala P. K., Bala M., Sinha A. K. and Dogra R.. "Radioprotective potential of an herbal extract of *Tinospora cordifolia*." *Journal of Radiation Research* 45, no. 1 (2004): 61–68.

Gupta `d Sharma V. "Ameliorative effects of *Tinospora cordifolia* root extract on histopathological and biochemical changes induced by aflatoxin-B1 in mice kidney." *Toxicology International* 18, no. 2 (2011): 94–98.

Iwasa K., Kim H-S., Wataya Y. and Lee D-U. "Antimalarial activity and structure-activity relationships of protoberberine alkaloids." *European Journal of Medicinal Chemistry* 33, no. 1 (1998): 65–69.

Jagetia G. C. and Baliga M. S. "Effect of Alstonia scholaris in enhancing the anticancer activity of berberine in the Ehrlich ascites carcinoma-bearing mice." *Journal of Medicinal Food* 7, no. 2 (2004): 235–244.

Jagetia G. C. and Rao S. K. "Evaluation of cytotoxic effects of dichloromethane extract of guduchi (*Tinospora cordifolia* Miers ex Hook F & THOMS) on cultured HeLa cells." *Evidence-Based Complementary and Alternative Medicine* 3, no. 2 (2006a): 267–272.

Jagetia G. C. and Rao S. K. "Evaluation of the antineoplastic activity of guduchi (*Tinospora cordifolia*) in Ehrlich ascites carcinoma bearing mice." *Biological and Pharmaceutical Bulletin* 29, no. 3 (2006b): 460–466.

Jahfar M. "Glycosyl composition of polysaccharide from *Tinospora cordifolia*." *Acta pharmaceutica (Zagreb, Croatia)* 53, no. 1 (2003): 65–69.

Jorge T. F., Mata A. T. and António C. "Mass spectrometry as a quantitative tool in plant metabolomics." *Philosophical Transactions of the Royal Society A: Mathematical, Physical and Engineering Sciences* 374, no. 2079 (2016): 20150370.

Kalikar M. V., Thawani V. R., Varadpande U. K., Sontakke S. D., Singh R. P. and Khiyani R. K. "Immunomodulatory effect of *Tinospora cordifolia* extract in human immuno-deficiency virus positive patients." *Indian Journal of Pharmacology* 40, no. 3 (2008): 107–110.

Kapil A. and Sharma S. "Immunopotentiating compounds from *Tinospora cordifolia*." *Journal of Ethnopharmacology* 58, no. 2 (1997): 89–95.

Kapur P., Wuttke W., Jarry H. and Seidlova-Wuttke D. "Beneficial effects of β-Ecdysone on the joint, epiphyseal cartilage tissue and trabecular bone in ovariectomized rats." *Phytomedicine* 17, no. 5 (2010): 350–355.

Khaleque A., Miah M. A. W., Huq M. S. and Abdul B. K. "*Tinospora cordifolia*. IV. Isolation of heptacosanol, β-sitosterol, and three other compounds tinosporidine, cordifol, and cordifolone." *Pakistan Journal of Scientific and Industrial Research* 14, no. 6 (1971): 481–483.

Khan, Md M., Sadul Haque M. and Chowdhury Md S. I. "Medicinal use of the unique plant *Tinospora cordifolia*: evidence from the traditional medicine and recent research." *Asian Journal of Medical and Biological Research* 2, no. 4 (2016): 508–512.

Khuda M. Q., Khaleque A., Basar K. A., Rouf M. A., Khan M. A. and Roy N. "Studies on *Tinospora cordifolia* II: Isolation of tinosporine, tinosporic acid and tinosporol from the fresh creeper, Sci Res, (Dacca), III (1966) 9." *Chemistry Abstracts* 65 (1966): 10549a.

Kohno H., Maeda M., Tanino M., Tsukio Y., Ueda N., Wada K., Sugie S., Mori H. and Tanaka T. "A bitter diterpenoid furanolactone columbin from *Calumbae radix* inhibits azoxymethane-induced rat colon carcinogenesis." *Cancer Letters* 183, no. 2 (2002): 131–139.

Lade S., Sikarwar P. S., Ansari, Md. A., Khatoon S., Kumar N., Yadav H. K. and Ranade S. A. "Diversity in a widely distributed dioecious medicinal plant, *Tinospora cordifolia* (Willd.) Miers. ex. Hook F. and Thomas." *Current Science* 114, no. 7 (2018): 1520–1526.

Leyon P. V. and Kuttan G. "Inhibitory effect of a polysaccharide from *Tinospora cordifolia* on experimental metastasis." *Journal of Ethnopharmacology* 90, no. 2–3 (2004): 233–237.

Maliyakkal N., Pai N. and Rangarajan A. Cytotoxic and apoptotic activities of extracts of *Withania somnifera* and *Tinospora cordifolia* in human breast cancer cells. *International Journal of Applied Research in Natural Products* 6, no. 4 (2013): 1–10.

Maurya R. and Handa S. S. "Tinocordifolin, a sesquiterpene from *Tinospora cordifolia.*" *Phytochemistry* 49, no. 5 (1998): 1343–1345.

Maurya R., Wazir V., Kapil A. and Kapil R. S. "Cordifoliosides A and B, two new phenylpropene disaccharides from *Tinospora cordifolia* possessing immunostimulant activity." *Natural Product Letters* 8, no. 1 (1996): 7–10.

Moody J. O., Robert V. A., Connolly J. D. and Houghton P. J. "Anti-inflammatory activities of the methanol extracts and an isolated furanoditerpene constituent of Sphenocentrum jollyanum Pierre (Menispermaceae)." *Journal of Ethnopharmacology* 104, no. 1–2 (2006): 87–91.

Naik D., Dandge C. and Rupanar S. "Determination of chemical composition and evaluation of antioxidant activity of essential oil from *Tinospora cordifolia* (Willd.) Leaf." *Journal of Essential Oil Bearing Plants* 17, no. 2 (2014): 228–236.

Nair P. K. R., Melnick S. J., Ramachandran R., Escalon E. and Ramachandran C. "Mechanism of macrophage activation by (1, 4)-α-D-glucan isolated from *Tinospora cordifolia.*" *International Immunopharmacology* 6, no. 12 (2006): 1815–1824.

Nikolić D., Gödecke T., Chen S. N., White J., Lankin D. C., Pauli G. F. and van Breemen R. B. "Mass spectrometric dereplication of nitrogen-containing constituents of black cohosh (Cimicifuga racemosa L.)". *Fitoterapia* 83, no. 3 (2012): 441–460.

OMS. WHO Traditional Medicine Strategy 2002–2005. WHO, 2002.

Paliwal R., Kumar R., Choudhury D. R., Singh A. K., Kumar S., Kumar A., Bhatt K. C., et al. "Development of genomic simple sequence repeats (g-SSR) markers in *Tinospora cordifolia* and their application in diversity analyses." *Plant Gene* 5 (2016): 118–125.

Pan, Li, Terrazas C., Lezama-Davila C. M., Rege N., Gallucci J. C., Satoskar A. R. and Kinghorn A. D.. "Cordifolide A, a sulfur-containing clerodane diterpene glycoside from *Tinospora cordifolia.*" *Organic Letters* 14, no. 8 (2012): 2118–2121.

Panchabhai T. S., Kulkarni U. P. and Rege N. N. "Validation of therapeutic claims of *Tinospora cordifolia*: a review." *Phytotherapy Research: An International Journal Devoted to Pharmacological and Toxicological Evaluation of Natural Product Derivatives* 22, no. 4 (2008): 425–441.

Patel M. B. and Mishra S. "Hypoglycemic activity of alkaloidal fraction of *Tinospora cordifolia*." *Phytomedicine* 18, no. 12 (2011): 1045–1052.

Peer F. and Sharma M. C. "Therapeutic evaluation of *Tinospora cordifolia* in CCl4 induced hepatopathy in goats." *Indian Journal of Veterinary Medicine* 9 (1989): 154–156.

Pendse G. P. and Bhatt S. K. "Chemical Examination of some Indian Medicinal Plants, *Tinospora cordifolia*, Solanum xanthocarpum and Fumaria officinalis in Bengal Pharmacopoea." *Indian Journal of Medical Research* 20 (1932): 663–670.

Pendse V. K., Dadhich A. P., Mathur P. N., Bal M. S. and Madan B. R. "Antiinflammatory, immunosuppressive and some related pharmacological actions of the water extract of Neem Giloe (*Tinospora cordifolia*): A preliminary report." *Indian Journal of Pharmacology* 9, no. 3 (1977): 221–224.

Pramanik A., Gangopadhyay M., Sharma B. D., Balakrishnan N. P., Rao R. R., Hajra P. K. Menispermaceae. In: *Flora of India*; Sharma B. D.; Balakrishnan N. P.; Rao R. R.; Hajra P. K., Eds.; Botanical Survey of India, Calcutta, India, 1993.

Pushp P., Sharma N., Joseph G. S. and Singh R. P. "Antioxidant activity and detection of (–) epicatechin in the methanolic extract of stem of *Tinospora cordifolia*." *Journal of Food Science and Technology* 50, no. 3 (2013): 567–572.

Qudrat-I-Khuda M., Khaleque A. and Ray N. "'*Tinospora cordifolia*.' L. Constituents of the plant fresh from the field." *Scientific Research (Dacca)* 1 (1964): 177–183.

Rana V., Thakur K., Sood R., Sharma V. and Sharma T. R. "Genetic diversity analysis of *Tinospora cordifolia* germplasm collected from northwestern Himalayan region of India." *Journal of Genetics* 91, no. 1 (2012): 99–103.

Reddy S. S., Ramatholisamma P., Karuna R. and Saralakumari D. "Preventive effect of *Tinospora cordifolia* against high-fructose diet-induced insulin resistance and oxidative stress in male Wistar rats." *Food and Chemical Toxicology* 47, no. 9 (2009): 2224–2229.

Reddy N. M. and Reddy R. N. "*Tinospora cordifolia* chemical constituents and medicinal properties: a review." *Scholars Academic Journal of* Pharmacy 4, no. 8 (2015): 364–369.

Reemtsma T. "Liquid chromatography–mass spectrometry and strategies for trace-level analysis of polar organic pollutants." *Journal of Chromatography A* 1000, no. 1–2 (2003): 477–501.

Rout S. P., Choudary K. A., Kar D. M., Das L. and Jain A. "Plants in traditional medicinal system-future source of new drugs." *International Journal of Pharmacy and Pharmaceutical Sciences* 1, no. 1 (2009): 1–23.

Saha S. and Ghosh S. "*Tinospora cordifolia*: One plant, many roles." *Ancient Science of Life* 31, no. 4 (2012): 151–159.

Samanani N. and Facchini P. J. "Isolation and partial characterization of norcoclaurine synthase, the first committed step in benzylisoquinoline alkaloid biosynthesis, from opium poppy." *Planta* 213, no. 6 (2001): 898–906.

Sarma D. N. K., Khosa R. L., Chansauria J. P. N. and Sahai M. "Antistress activity of *Tinospora cordifolia* and Centella asiatica extracts." *Phytotherapy Research* 10, no. 2 (1996): 181–183.

Sarma D. N. K., Padma P. and Khosa R. L. "Constituents of *Tinospora cordifolia* root." *Fitoterapia* 69, no. 6 (1998): 541–542.

Sarma D. N. K., Sameksha K. and Khosa R. L. "Alkaloids from *Tinospora cordifolia* miers." *Journal of Pharmaceutical Sciences and Research* 1, no. 1 (2009): 26–27.

Sarma D. N. K., Khosa R. L., Chansauria J. P. N. and Sahai M. "Antiulcer activity of *Tinospora cordifolia* Miers and Centella asiatica Linn extracts." *Phytotherapy Research* 9, no. 8 (1995): 589–590.

Seidlova-Wuttke D., Christel D., Kapur P., Nguyen Ba T., Jarry H. and Wuttke W. "β-Ecdysone has bone protective but no estrogenic effects in ovariectomized rats." *Phytomedicine* 17, no. 11 (2010): 884–889.

Sengupta S., Mukherjee A., Goswami R. and Basu S. "Hypoglycemic activity of the antioxidant saponarin, characterized as α-glucosidase inhibitor present in *Tinospora cordifolia*." *Journal of Enzyme Inhibition and Medicinal Chemistry* 24, no. 3 (2009): 684–690.

Murthi K. R. S. *Sharangdhar-Samhita*. Translated by K. R. Shrikanta Murthi; Chaukhamba Orientalia, Varanasi, India, 1984.

Sharma A., Gupta A., Singh S. and Batra A. "*Tinospora cordifolia* (Willd.) Hook. F. & Thomson-A plant with immense economic potential." *Journal of Chemical and Pharmaceutical Research* 2, no. 5 (2010a): 327–333.

Sharma U., Bala M., Kumar P., Rampal G., Kumar N., Singh B. and Arora S.. "Antimutagenic extract from *Tinospora cordifolia* and its chemical composition." *Journal of Medicinal Plants Research* 4, no. 23 (2010b): 2488–2494.

Sharma R., Amin H. and Prajapati P. K. "Antidiabetic claims of Tinospora cordifolia (Willd.) Miers: critical appraisal and role in therapy." *Asian Pacific Journal of Tropical Biomedicine* 5, no. 1 (2015): 68–78.

Singh N., Singh S. M. and Shrivastava P. "Immunomodulatory effect of *Tinospora cordifolia* in tumor-bearing host." *Oriental Pharmacy and Experimental Medicine* 3, no. 2 (2003a): 72–79.

Singh S. S., Pandey S. C., Srivastava S., Gupta V. S., Patro B. and Ghosh A. C. "Chemistry and medicinal properties of *Tinospora cordifolia* (Guduchi)." *Indian Journal of Pharmacology* 35, no. 2 (2003b): 83–91.

Singh K., Kadyan S., Panghal M. and Yadav J. P. "Assessment of genetic diversity in *Tinospora cordifolia* by inter simple sequence repeats (ISSR) and expressed sequence tagged-simple sequence repeats (EST-SSR)." *TC* 6, no. 10 (2014): 520–524.

Singh K. J. and Thakur A. K. "Medicinal plants of the Shimla hills, Himachal Pradesh: a survey." *International Journal of Herbal Medicine* 2, no. 2 (2014): 118–127.

Singh D. and Chaudhuri P. K. "(+) Corydine from the Stems of *Tinospora cordifolia*." *Asian Journal of Chemistry* 27, no. 4 (2015): 1567–1568.

Singh D. and Chaudhuri P. K. "Chemistry and Pharmacology of *Tinospora cordifolia*." *Natural Product Communications* 12, no. 2 (2017): 299–308.

Sivasubramanian A., Narasimha K. K. G., Rathnasamy R. and Campos A. M. F. O. "A new antifeedant clerodane diterpenoid from *Tinospora cordifolia*." *Natural Product Research* 27, no. 16 (2013): 1431–1436.

Stanely P., Prince M. and Menon V. P. "Hypoglycaemic and other related actions of Tinospora cordifolia roots in alloxan-induced diabetic rats." *Journal of Ethnopharmacology* 70, no. 1 (2000): 9–15.

Upadhyay A. K., Kumar K., Kumar A. and Mishra H. S. "*Tinospora cordifolia* (Willd.) Hook. f. and Thoms. (Guduchi)–validation of the Ayurvedic pharmacology through experimental and clinical studies." *International Journal of Ayurveda Research* 1, no. 2 (2010): 112–121.

Van Kiem P., Van Minh C., Dat N. T., Hang D. T., Nam N. H., Cuong N. X., Huong H. T. and Lau T. V. "Aporphine alkaloids, clerodane diterpenes, and other constituents from *Tinospora cordifolia*." *Fitoterapia* 81, no. 6 (2010): 485–489.

Wadood N., Wadood A. and Shah S. A. W. "Effect of *Tinospora cordifolia* on blood glucose and total lipid levels of normal and alloxan-diabetic rabbits." *Planta Medica* 58, no. 2 (1992): 131–136.

Warrier P. K., Nambiar V. P. K. and Ramankutty C. (eds.), *Indian Medicinal Plants, A Compendium of 500 Species*. (5 vols.) Orient Longman, New Delhi, India, 1993–1996.

Index